Edexcel GCSE

Design and Technology

Resistant Materials

Student Book

Barry Lambert

A PEARSON COMPANY

Acknowledgements

Published by Pearson Education Limited, a company incorporated in England and Wales, having its registered office at Edinburgh Gate, Harlow, Essex, CM20 2JE. Registered company number: 872828

Edexcel is a registered trademark of Edexcel Limited

Text © Barry Lambert 2010

First published 2010

12 11
10 9 8 7 6 5 4 3

British Library Cataloguing in Publication Data

A catalogue record for this book is available from the British Library

ISBN 978 1 846907 55 5

Copyright notice

Edited by Helen Gough and Sarah Christopher
Designed by Oxford Designers and Illustrators 2010
Original illustrations © Pearson Education Limited 2010
Illustrated by Oxford Designers and Illustrators 2010
Printed and bound in China (GCC)

Acknowledgements

The author would like to thank Cheltenham College for their continued support during the writing of this book. He would also like to thank the pupils for allowing their work to be used, especially Will Ripley and Mark Doumler. A special thank you to Michelle Watling and Charlie McKegney for allowing their work to be included too.

Also to Sue, James and Harriet for their long standing patience and love.

Picture Credits

The publisher would like to thank the following for their kind permission to reproduce their photographs:

Alamy Images: Nigel Cattlin 88, Greenshoots Communications 113, Phil Degginger 30, Sundlof - EDCO 63, David J. Green - industry 108, Graeme Peacock 19, Iain Davidson Photographic 53, Ian Miles-Flashpoint Pictures 82, Shenval 18; **DK Images:** Michael Crockett 77; **Getty Images:** Dave M. Benett 121, Fisher / Thatcher 105, Christopher Furlong 116, Jun Sato / Wireimage 27/3; **iStockphoto:** 21, darren baker 104, chang 94, Alexey Dudoladov 106, Joseph Gareri 116/2, David Gilder 102, Abel Leão 114/2, Hans F. Meier 117, Michael Sleigh 22, Denis Jr. Tangney 12, Sining Zhang 87; **Pearson Education Ltd:** Trevor Clifford 5, 15, 15/2, 15/3, 15/4, 15/5, 16, 16/2, 17, 17/2, 17/3, 17/4, 34, 34/2, 35, 35/2, 35/3, 35/4, 36, 37, 37/2, 37/3, 38, 38/2, 38/3, 38/4, 38/5, 39, 40, 40/2, 40/3, 41, 41/2, 42, 42/2, 42/3, 43, 43, 43/2, 43/3, 43/4, 46, 52, 52/2, 52/3, 62, 70, 73, 76, 77/2, 86, 99, **Photolibrary.com:** Matt Meadows / Peter Arnold Images 47; **Rex Features:** David Cairns 114, Action Press 48, Sipa Press 79; **Science Photo Library Ltd:** MARK CLARKE 28, BRUCE FRISCH 26; **Shutterstock:** 27, 59, 113/2, Racheal Grazias 27/2, george green 112, Chad Gordon Higgins 25, Hugo Maes 120, paul prescott 61; **SuperStock:** Bill Barley 107

Cover images: *Front:* **iStockphoto:** Marek Kosmal

All other images © Pearson Education

We are grateful to the following for permission to reproduce copyright material:

Logos

Logo on page 104 from www.bluetooth.com, © Bluetooth SIG

Photos

Photos on page 29, UV photochromic paint from http://www.behance.net/Gallery/Fairytale-Catering-identity/131729; Photos on page 30, switchable Smart-tint glass technology from www.reflexglass.co.uk, Smart-Tint™

Every effort has been made to trace the copyright holders and we apologise in advance for any unintentional omissions. We would be pleased to insert the appropriate acknowledgement in any subsequent edition of this publication.

The websites used in this book were correct and up to date at the time of publication. It is essential for tutors to preview each website before using it in class so as to ensure that the URL is still accurate, relevant and appropriate. We suggest that tutors bookmark useful websites and consider enabling students to access them through the school/college intranet.

Disclaimer

Contents: delivering the Edexcel GCSE Resistant Materials specification

Welcome to Edexcel GCSE Resistant Materials

Why should I choose GCSE Resistant Materials?

Because you will:

- think creatively
- solve problems
- design your own product of the future
- make models
- test your ideas.

What will I learn?

GCSE Resistant Materials Technology (RMT) covers a wide range of activities based on designing and making products that are manufactured using materials such as wood, metal and plastics in many forms. As well as learning hand skills, you will use a range of industrial processes to shape and form materials into functioning products.

Over the course of two years you will develop a whole range of creative designing and making skills, technical knowledge and understanding relating to RMT, and invaluable transferable skills such as problem solving and time management.

Unit 1: Your controlled assessment

This is the unit where you can really get stuck in! You can either create a combined design and make activity, where you design a product and then make a model of it, or you can complete a separate design and make activity where you design one product and make another.

Unit 2: Knowledge and Understanding of Resistant Materials

This unit focuses on developing your knowledge and understanding of a wide range of materials and processes used in the field of design and technology. You will learn about industrial and commercial practices and the importance of quality checks, and the health and safety issues that have to be considered. What you learn in this unit will be applied during Unit 1 Creative Design and Make Activities.

How will I be assessed?

Unit 1: Creative Design and Make Activities is the controlled assessment (coursework) unit. This means that your work will be internally assessed by your teacher. It is worth **60%** of your overall course.

For Unit 2: Knowledge and Understanding of Resistant Materials, you will sit a 1 hour 30 minute examination that is assessed by Edexcel. It is worth **40%** of your overall course.

The great thing about the course is that each of the units can be re-taken once. This means that if you don't achieve the mark you wanted, then you can have another go!

What can I do after I've completed the course?

Many students have enjoyed studying GCSE RMT so much that they go on to study A Level Product Design: RMT for a further two years. However, it is possible to study any D&T related course at post-16.

Creative students usually study one or more of the creative subjects including, A Level Art and Design, Media and/or Film, BTEC National Diplomas in Art and Design or Media and the 14-19 Diploma in Creative and Media. Of course, if post-16 is not for you, employers value this GCSE RMT qualification as it develops creative, technical and transferable skills.

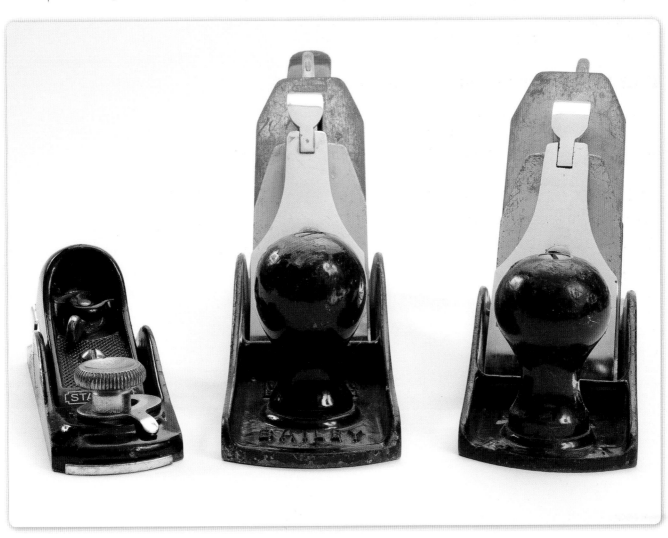

About this book

Objectives provide a **clear overview** of what you will learn in the section.

Engaging photos **bring resistant materials** to life.

Chapter 1 Materials and components
Woods

Objectives

- **Describe** the aesthetics and properties of hardwoods and softwoods.

- **Explain** their uses, advantages and disadvantages when manufacturing products.

The types of woods

Wood is a very versatile material. It has been used throughout history for many purposes such as fuel for fires, weapons for hunting, cooking utensils, transportation, structures for houses and furniture. Its fibres are also used in the production of paper.

There are two main types of timber (wood from different trees):

- hardwoods
- softwoods.

Figure 1.1: Bare trees in winter are hardwoods; those that keep their leaves are usually softwoods

Hardwoods

The term hardwood is associated with timbers coming from deciduous (leaf-losing) trees. The use of the word hardwood is not always a true description of the wood's hardness. Some hardwoods such as balsa wood are much lighter in weight and softer than most softwoods. Hardwoods are generally slower growing, making them denser. Some trees take up to 100 years to reach full maturity, and this makes them very expensive to buy. The hardwoods you will be required to know are:

- oak
- mahogany
- beech
- ash.

The aesthetics of hardwoods

The colour of timbers varies enormously. Oak varies in colour depending upon whether it comes from the USA or Europe. Generally oak is described as being pale brown in colour with distinctive growth rings. Mahogany is a reddish-brown colour. Beech varies but is often a light brown colour whereas ash is a creamy white colour.

ResultsPlus
Exam Question Report

Give *three* properties of oak.
(3 marks, 2008)

How students answered
A good number of students scored well on this section since they knew about specific properties of oak and were able to recall them.

Property 1
'Hard' was the most popular answer here

38%	1 mark

Property 2
'Tough' was a popular second response for many candidates.

39%	1 mark

Property 3
Naturally a third property was going to be difficult for many candidates if they had already scored two marks for the first two properties named, but 'durable' and 'dense' were answers produced by a good number of candidates.

30%	1 mark

ResultsPlus features combine real exam performance data with examiner insight to give **guidance on how to achieve better results**.

Key terms are highlighted in the text with a full definition given in the glossary at the end of the book to enable you to develop your understanding of Resistant Materials terminology

13

Most woods are easily identifiable by their colour and their grain patterns. As a result they are carefully chosen for specific applications. Sometimes, for example, oak is cut in a particular way to expose a particular grain pattern or 'figure' where gold- and silver-coloured slivers can be seen. This makes it very desirable and it is often used in high-quality furniture. It does also make it very expensive.

Wood is an organic material and will decay over time. In wetter conditions this decay will be quicker. It is also subject to conditions such as dry rot and attack by pests which make it structurally weaker. Natural wood is also subject to movements such as warping, bowing, cupping and splitting (see Figure 1.2). All of these will in some cases make the piece of wood unusable.

Wood also contains knots, which are formed where branches grow from the main trunk or where a bud was formed. Knots will generate weaknesses in the timber but may in some cases be used from an aesthetic point of view.

Support Activity

Make a list of five differences between hardwoods and softwoods.

Stretch and Support Activities provide extra support to ensure understanding and opportunities to stretch your knowledge.

Stretch Activity

Some hardwoods do not lose their leaves in winter. Find out which hardwood trees do not lose their leaves in winter.

Quick notes summarise content into manageable chunks.

Quick notes
Hardwoods come from deciduous trees that can take over 100 years to grow. Oak, mahogany, beech and ash are examples of hardwood and are typically dense and tough.

Softwoods come from coniferous trees that mature within approximately thirty years and so is often cheaper than hardwood. Pine is an example of a softwood and is typically less dense than hardwood.

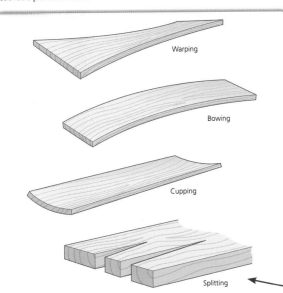

Figure 1.2: Defects in natural wood: warping, bowing, cupping and splitting

The properties of hardwoods
Hardwoods contain much more fibrous material than softwoods. The fibres are smaller and more compact which gives the wood greater mechanical strength and hardness. In general, the greater the density of the wood, the greater its mechanical strength. Mahogany, deemed to be a medium-dense hardwood, is excellent for furniture because of its colour, texture and straight, even grain. Balsa wood, which is useful for model aircraft building and modelling, is very light and not very dense.

Clear and accessible diagrams **highlight key concepts.**

Apply It!

If necessary in Unit 1: Make activity, use a plane in the direction of the grain. When planing end grain, use a piece of scrap wood to prevent splitting, or plane into the centre from both sides. Figure 2.8 shows how to do this.

Apply it! helps you relate key content to the controlled assessment activities.

examzone

Exam Zone is a dedicated suite of revision resources for **complete exam success.**

We've broken down the six stages of revision to ensure that you are prepared every step of the way.

Zone In: How to get into the perfect 'zone' for your revision.

Planning Zone: Tips and advice on how to effectively plan your revision.

Know Zone: All the facts you need to know and practice exam questions at the end of every chapter.

Chapter overview: Outlines the key issue that the chapter examines. Keep this issue in mind as you work through the Know Zone pages.

Key terms: Lists the important Resistant Materials terminology and provides a matching exercise to ensure that you can understand and apply the words.

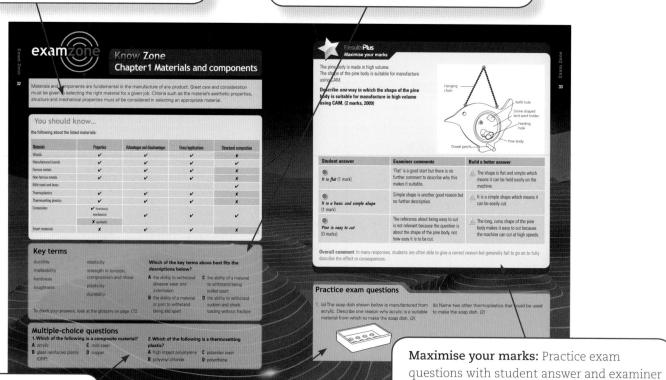

Multiple-choice questions: Provides a number of questions to test your understanding and an opportunity to get to grips with the question type.

Practice exam questions: Test yourself on the chapter content with some exam-type questions.

Maximise your marks: Practice exam questions with student answer and examiner commentary (see next page).

Don't Panic Zone: Last-minute revision tips for just before the exam.

Exam Zone: An explanation of the assessment objectives, plus a chance to see what a real exam paper might look like.

Zone Out: What do you do after your exam? This section contains information on how to get your results and answers to frequently asked questions on what to do next.

ResultsPlus

These features are based on the actual marks that students have achieved in past exams. They are combined with expert advice and guidance from examiners to show you **how to achieve better results**.

There are five different types of ResultsPlus feature throughout this book:

ResultsPlus — Exam Question Report

The handle of a garden spade is made of ABS plastic and is formed using the injection-moulding process. Explain one reason why ABS plastic needs the property of plasticity. (2 marks, 2007)

How students answered

The overwhelming majority failed to make any comment about plasticity or to give any form of explanation about why ABS needs the property of plasticity.

70%   0 marks)

Some comments made about the fact that the plastic needs to be squeezed into the mould but with little or no explanation as to why.

28%   1 mark

Good comment about plasticity and its relevance to being easily injected under pressure into the mould.

2%   2 marks

ResultsPlus — Build Better Answers

Explain **two** reasons why manufacturers create 3D 'virtual' products on screen. (4 marks)

■ **Basic answers (0–1 mark)**
Make general statements about manufacturers creating virtual products but without offering any explanation.

● **Good answers (1–2 marks)**
Explain one advantage, such as changes can be made easily because the work and files are stored on a disc.

▲ **Excellent answers (3–4 marks)**
Explain two advantages, such as slower growing so you can see the whole product from all angles and tool paths routes can be assessed and therefore clamping points and manufacturing times can be evaluated.

ResultsPlus — Exam Tip

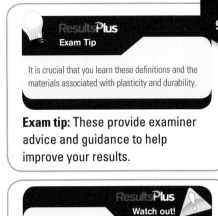

It is crucial that you learn these definitions and the materials associated with plasticity and durability.

Exam tip: These provide examiner advice and guidance to help improve your results.

ResultsPlus — Watch out!

Be careful not to confuse scribers and centre punches. You hit a centre punch with a hammer; if you do this to a scriber, you will blunt or break its tip. However, if you try to use a punch for marking out, it will not be as accurate.

Watch out! These warn you about common mistakes and misconceptions that examiners frequently see students make. Make sure that you don't repeat them!

Exam Question Report These show previous exam questions with details about how well students answered them.

- **Red** shows the number of students who scored low marks (less than 35% of the total marks)
- **Orange** shows the number of students who did okay (scoring between 35% and 70% of the total marks)
- **Green** shows the number of students who did well (scoring over 70% of the total marks).

They explain how students could have achieved the top marks so that you can make sure that you answer these questions correctly in future.

Build Better Answers These give you an opportunity to answer some exam-style questions. They contain tips for what a basic ■ , good ● and excellent ▲ answer will contain.

ResultsPlus — Maximise your marks

The pine body is made in high volume. The shape of the pine body is suitable for manufacture using CAM.

Describe *one* way in which the shape of the pine body is suitable for manufacture in high volume using CAM. (2 marks, 2009)

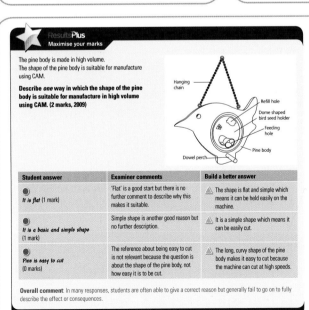

Student answer	Examiner comments	Build a better answer
● It is flat (1 mark)	'Flat' is a good start but there is no further comment to describe why this makes it suitable.	▲ The shape is flat and simple which means it can be held easily on the machine.
● It is a basic and simple shape (1 mark)	Simple shape is another good reason but no further description.	▲ It is a simple shape which means it can be easily cut.
● Pine is easy to cut (0 marks)	The reference about being easy to cut is not relevant because the question is about the shape of the pine body, not how easy it is to be cut.	▲ The long, curvy shape of the pine body makes it easy to cut because the machine can cut at high speeds.

Overall comment: In many responses, students are often able to give a correct reason but generally fail to go on to fully describe the effect or consequences.

Maximise your marks These are featured in the Know Zone at the end of each chapter. They include an exam-style question with a student answer, examiner comments and an improved answer so that you can see how to build a better response. Follow the green triangles to create a full-mark answer.

Knowledge and Understanding of Resistant Materials: introduction

Unit 2: Knowledge and Understanding of Resistant Materials This unit requires you to demonstrate your knowledge and understanding of eight areas related to resistant materials in a 1½ hour exam. Each area is presented as a chapter.

Chapter 1: Materials and components You need to understand a wide range of materials and components to make informed choices about their use in products or components.

Chapter 2: Tools and equipment You need to be able to identify a variety of tools and pieces of equipment used for marking out and measuring and tools related to various wasting processes. You should be able to name them and describe their uses and applications.

Chapter 3: Industrial and commercial processes You need to be able to show an awareness of the scale of production, materials processing, joining methods, adhesives, heat treatment, finishing techniques, jigs and patterns and health and safety so that when you design and make products you are aware of what can be done with materials.

Chapter 4: Analysing products Products surround us and we use and interact with them in all that we do, so you should be able to make constructive comments about products and to make comparisons between similar products using a set of specific criteria.

Chapter 5: Designing products This is one of the reasons why you are probably studying Design and Technology and you should be able to respond creatively and imaginatively to a set design brief and given list of specification points.

Chapter 6: Technology Technology is developing at an ever-increasing pace and you should be aware of the impact ICT is having on society and the way in which digital media and new technologies are developing.

Chapter 7: Sustainability Sustainable design involves using energy and materials in such ways as to minimise the depletion of finite resources and to reduce waste and pollution.

Chapter 8: Ethical design and manufacture In today's society consumerism has created an ever-increasing level of demand for new products and designers must consider ethical, moral and cultural issues when looking for new ideas and products.

How much is it worth? Unit 2: Knowledge and Understanding of Resistant Materials Technology is worth **40%** of your overall GCSE Design & Technology Resistant Materials course. This is an important exam, so make sure that you are familiar with its structure, take time to revise and practice your exam technique.

Chapter 1 Materials and components Woods

Objectives

- **Describe** the aesthetics and properties of hardwoods and softwoods.

- **Explain** their uses, advantages and disadvantages when manufacturing products.

ResultsPlus
Exam Question Report

Give *three* properties of oak. (3 marks, 2008)

How students answered

A good number of students scored well on this section since they knew about specific properties of oak and were able to recall them.

Property 1
'Hard' was the most popular answer here

38% got 1 mark

Property 2
'Tough' was a popular second response for many candidates.

39% got 1 mark

Property 3
Naturally a third property was going to be difficult for many candidates if they had already scored two marks for the first two properties named, but 'durable' and 'dense' were answers produced by a good number of candidates.

30% got 1 mark

The types of woods

Wood is a very versatile material. It has been used throughout history for many purposes such as fuel for fires, weapons for hunting, cooking utensils, transportation, structures for houses and furniture. Its fibres are also used in the production of paper.

There are two main types of timber (wood from different trees):

- hardwoods
- softwoods.

Figure 1.1: Bare trees in winter are hardwoods; those that keep their leaves are usually softwoods

Hardwoods

The term hardwood is associated with timbers coming from deciduous (leaf-losing) trees. The use of the word hardwood is not always a true description of the wood's hardness. Some hardwoods such as balsa wood are much lighter in weight and softer than most softwoods. Hardwoods are generally slower growing, making them denser. Some trees take up to 100 years to reach full maturity, and this makes them very expensive to buy. The hardwoods you will be required to know are:

- oak
- mahogany
- beech
- ash.

The aesthetics of hardwoods

The colour of timbers varies enormously. Oak varies in colour depending upon whether it comes from the USA or Europe. Generally oak is described as being pale brown in colour with distinctive growth rings. Mahogany is a reddish-brown colour. Beech varies but is often a light brown colour whereas ash is a creamy white colour.

Most woods are easily identifiable by their colour and their grain patterns. As a result they are carefully chosen for specific applications. Sometimes, for example, oak is cut in a particular way to expose a particular grain pattern or 'figure' where gold- and silver-coloured slivers can be seen. This makes it very desirable and it is often used in high-quality furniture. It does also make it very expensive.

Wood is an organic material and will decay over time. In wetter conditions this decay will be quicker. It is also subject to conditions such as dry rot and attack by pests which make it structurally weaker. Natural wood is also subject to movements such as warping, bowing, cupping and splitting (see Figure 1.2). All of these will in some cases make the piece of wood unusable.

Wood also contains knots, which are formed where branches grow from the main trunk or where a bud was formed. Knots will generate weaknesses in the timber but may in some cases be used from an aesthetic point of view.

Support Activity

Make a list of five differences between hardwoods and softwoods.

Stretch Activity

Some hardwoods do not lose their leaves in winter. Find out which hardwood trees do not lose their leaves in winter.

Quick notes

Hardwoods come from deciduous trees that can take over 100 years to grow. Oak, mahogany, beech and ash are examples of hardwood and are typically dense and tough.

Softwoods come from coniferous trees that mature within approximately thirty years and so is often cheaper than hardwood. Pine is an example of a softwood and is typically less dense than hardwood.

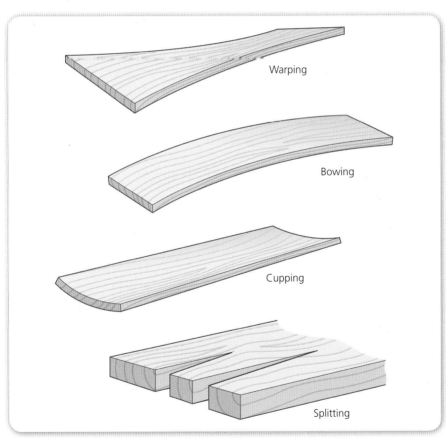

Figure 1.2: Defects in natural wood: warping, bowing, cupping and splitting

The properties of hardwoods

Hardwoods contain much more fibrous material than softwoods. The fibres are smaller and more compact which gives the wood greater mechanical strength and hardness. In general, the greater the density of the wood, the greater its mechanical strength. Mahogany, deemed to be a medium-dense hardwood, is excellent for furniture because of its colour, texture and straight, even grain. Balsa wood, which is useful for model aircraft building and modelling, is very light and not very dense.

The hardwood sill is made from mahogany. Explain **two** advantages of using mahogany rather than pine, a softwood, for the sill. (4 marks, 2007)

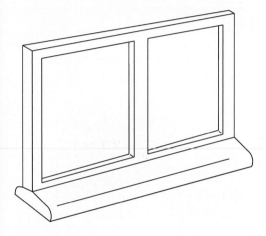

■ **Basic answers (0–1 marks)**
Make general statements about the difference between hardwoods and softwoods without offering any detail about the advantages.

● **Good answers (2 marks)**
Identify one advantage, such as it being more durable, that might lead on to it lasting longer than pine or that it requires less maintenance.

▲ **Excellent answers (3–4 marks)**
Two advantages are given and fully explained, such as slower growing or more dense meaning it will last longer and harder, tougher which means it is more durable.

Softwoods

Softwoods are commonly classified as coniferous (cone-bearing). Because they are evergreen, they can reach maturity in about 30 years, making them cheaper than hardwoods and more sustainable. The only softwood you need to be familiar with is pine.

The aesthetics of softwood

Softwoods such as pine are very resinous and at times this resin will leak from the timber. Resin is very sticky and messy and will also come through a painted surface to make a horrible stain. Pine varies enormously in colour and will change colour when left exposed to sunlight for prolonged periods. Generally it is pale yellow in colour with brown streaks. Softwoods are also prone to decaying and warping, bowing, cupping and splitting.

The properties of softwoods

The structure of softwoods is generally made up from tube-like cells. This normally makes softwoods less dense than hardwoods. Softwoods are also more prone to water damage. This is because the timber absorbs the water like a sponge if the end grain is exposed and left untreated.

The growth and production of woods

During the growing season, normally spring through to the autumn, the tree's girth (thickness) increases along with its height. In most trees, the cells produced during the drier summer months have thicker cell walls. This summer growth is mainly responsible for the mechanical strength of the timber, and the variations in the look of the cells can be seen as annual rings. The annual rings can be counted to determine how old the tree is. Scientists use this information about the annual rings and growth patterns to give an indication of the climate and environmental conditions at the time of growth.

Once a tree has been felled, cut down and taken to a sawmill, it is converted ready for seasoning. After the timber has dried out, it is cut into smaller sections of common sizes and shapes. Most timber now sold in DIY stores is 'planed all round' (PAR), that is, each side has been planed.

Wood and type	Grain pattern	Properties	Uses	Advantages	Disadvantages
oak hardwood		hard tough durable high density	high-quality furniture garden benches boat building veneers	finishes well	contains an acid which corrodes steel
mahogany hardwood		durable medium density	indoor furniture interior woodwork window frames veneers	finishes well relatively easy to work	prone to warping some tropical types can be a bit soft and fibrous
beech hardwood		hard tough	workshop benches children's toys interior furniture kitchen chopping boards and worktops	turns well finishes well	prone to warping
ash hardwood		tough flexible good elasticity	sports equipment ladders laminated furniture tool handles	flexible	can become a bit splintered
pine softwood		lightweight	constructional woodwork (joists, roof trusses) floorboards children's toys garden decking	nice colour and grain pattern grows relatively quickly in comparison to hardwoods	prone to warping knots can fall out and leave holes

Table 1.1: Properties and uses of woods

Manufactured boards

Manufactured boards are man-made. Waste wood is used to make MDF and chipboard. Other manufactured boards, such as plywood, are layers of veneers glued together. Solid timber is used much less in the furniture and construction industries today than manufactured boards. The mass production of furniture is almost entirely based around a range of manufactured boards. Table 1.2 shows the advantages and disadvantages of manufactured boards.

Figure 1.3: Manufactured boards

Figure 1.4: Iron-on edging being applied to an exposed edge on a MDF board

The wine rack in Figure 1.5 has been manufactured from plywood. Thin sheets are very flexible and have been laminated over a former. As the glue has dried, they take a permanent form, which is then joined together to form the wine rack.

Figure 1.5: A plywood wine rack

Advantages	Disadvantages
• available in large flat sheets – 2440 × 1220 mm so can be used for large pieces of furniture without having to join pieces together	• sharp tools required when cutting manufactured boards, and tools easily blunted
• good dimensional stability – do not warp as much as natural timbers	• thin sheets do not stay flat and will bow unless supported
• can be decorated in a number of ways, e.g. with veneers or paint	• difficult to join in comparison with traditional construction methods – you cannot cut traditional woodwork construction joints such as finger or dovetail joints
• sheets of plywood and MDF are flexible and easy to bend over formers for laminating	• cutting and sanding some types of board generates hazardous dust particles
• waste from wood production can be used in making MDF, chipboard and hardboard	• edges must be treated and covered to hide unsightly edges and to stop water getting in, a process called concealing edges; this also helps to create an appearance of a solid piece of timber

Table 1.2: Advantages and disadvantages of manufactured boards

Board type	Aesthetics	Properties	Uses	Advantages	Disadvantages
plywood	made of layers (veneers) normally 1.5mm thick grain of each layer is at right angles to the layer either side of it, and there is an odd number so that the two outside layers run in same direction birch veneers frequently used on the outside layers resulting in an attractive surface	very strong in all directions resistant to splitting because layers are in alternate directions	boat building (exterior quality plywood) drawer bottoms and wardrobe bottoms tea chests cheaper grades used in construction industry for hoardings and shuttering	available in large sheets thicker sheets will not warp or twist thin sections can be laminated to create 2D shapes	thin sheets very flexible and will warp if not correctly stacked or supported
chipboard	no grain patterns surface often veneered or covered with a plastic laminate	made from waste products bonded together using very strong resins strong in all directions although not as strong as plywood not very resistant to water but moisture-resistant grades available	large floor boards and decking for loft spaces shelving kitchen worktops flat-packed furniture	makes good use of waste materials that are chipped up	not very good around water because it will just soak it up will chip and flake on edges if not protected
medium density fibreboard (MDF)	excellent surface finish, which can be veneered or painted	very dense stable and not affected by changing humidity levels will break down and absorb water if it gets very wet	flat-packed furniture drawer bottoms kitchen units heat and sound insulation	thin sheets can be formed to make 2D shapes	not very good with water because it will soak it up at the edges
hardboard	side very smooth and underside textured	made from compressed fibres that have been soaked in resin before being compressed	drawer bottoms cabinet backs smoothing out uneven floors lightweight internal door cladding	cheapest of manufactured boards	not very strong as it has no grain

Table 1.3: Properties and uses of manufactured boards

Metals

Metals

Objectives

- **Describe** the aesthetics and properties of ferrous and non-ferrous metals.

- **Understand** the composition of mild steel and brass.

- **Explain** their uses, advantages and disadvantages when manufacturing products.

- **Understand** and describe a selection of properties when selecting and using metals in product manufacture.

The types of metals

There are three main groups of metals:

- ferrous
- non-ferrous
- alloys.

Metals and metal alloys have many varied applications depending on their properties. Some metals are very good at carrying and supporting large loads, such as the girders that are used in constructing high-rise buildings or large factory warehouses, as in Figure 1.6.

Ferrous

This group of metals is composed mainly of ferrite or iron. Small amounts of other elements are added in their production, such as carbon, tungsten, chromium and nickel. Almost all ferrous metals are magnetic. Some examples of ferrous metals are mild steel, stainless steel and carbon steel.

Steel is generally a dull grey colour but some steels are black with a rough, pitted surface due to the processes used during their production. They will develop a surface oxide over time, which is a reddish-brown. Artists and sculptures have exploited this reaction to create some stunning sculptures that have oxidised over time. The 'Angel of the North' (Figure 1.7) is an excellent example of what happens to a steel structure over time, although a small amount of copper was alloyed with the steel to create a reddish appearance right from the start.

Figure 1.6: Steel girders used in the construction industry

The properties of metals

The useful properties of ferrous and non-ferrous metals are most commonly related to their strength, and this is classified in terms of tensile, compressive and shear strength.

Metals exhibit a range of different properties with some metals being harder than others, some more ductile than others and some more malleable. These properties are explained in detail on page 21.

Materials are chosen for specific products or components depending on what properties they have and how they will withstand being treated in particular ways, such as being pulled or squashed, for example. Some of the most common properties and uses of ferrous metals are shown in Table 1.4.

Figure 1.7: The 'Angel of the North' sculpture, which is made from steel

Metal	Properties	Uses	Advantages	Disadvantages
mild steel	tough malleable magnetic	structural steel girders car body panels	easily worked and joined, even in a school workshop relatively cheap widely available in numerous forms and sections can be recycled	will oxidise (rust) if left unprotected can only be case-hardened
stainless steel	hard tough excellent corrosion resistance	cutlery kitchen sinks pots and pans surgical instruments	easily cleaned does not need any surface finishing can be recycled high-lustre finish	difficult to use and join in a school workshop specialist welding equipment required for joining
carbon steel	ductile	nails, screws, nuts and general ironmongery	can be recycled	will oxidise (rust) if left unprotected can be easily heat-treated

Table 1.4: Properties and uses of ferrous metals

Stretch Activity

Research how a tiny change in the amount of carbon affects the hardness of the steel.

ResultsPlus
Exam Question Report

Give *two* properties that the brass pins must have.
(2 marks, 2009)

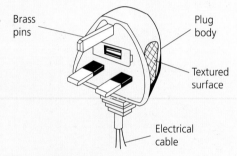

Brass pins

Plug body

Textured surface

Electrical cable

How students answered

First property
Often 'strong' was given as a property which is not specific enough.

48%   0 marks

A good number correctly responded with an appropriate property here such as hardness.

52%   1 mark

Second property
Naturally it is a little more difficult to come up with a second property but this part was poorly done as it was either left unanswered or a property such as, it will not rust, or, it is not magnetic, was given.

74%   0 marks

Toughness was quite a popular answer along with ductility and malleability.

26%   1 mark

Non-ferrous

Non-ferrous metals contain no iron and consist almost entirely of pure metals. Non-ferrous metals are not magnetic. Aluminium, copper, zinc and brass are all non-ferrous metals.

Aluminium has a reflective surface which can be polished to make it even more shiny and aesthetically appealing. Copper is a material that, over time, will develop a blue-green surface oxide. In some instances, houses are built with copper-clad roofs and rainwater guttering and downpipes, on the basis that, over a period of time, they will change colour. Brass is a goldy-yellow colour and has excellent resistance to corrosion. Zinc is a dull, blue-grey colour and when used for galvanising has a scaly appearance.

Non-ferrous metals are good conductors of heat and electricity. Electrical power distribution grids, such as the National Grid, rely on overhead cables made from steel-reinforced aluminium, attached to pylons, to deliver electricity to homes and businesses across the country. Aluminium is used in preference to copper because it is cheaper and has better resistance to weight characteristics. Home electrical systems rely on copper cabling to carry electricity around the house for lighting, water heating and power. Copper is used rather than aluminium because it can be made thinner than aluminium, enabling it to fit into smaller ducts and bend around corners more easily, while carrying the same power.

Metal	Properties	Uses	Advantages	Disadvantages
aluminium	lightweight soft ductile malleable good conductor of heat and electricity good corrosion resistance	window frames soft drink cans kitchen foil used in alloys	easily drawn into thin wires and sheets can be recycled easily cast	expensive difficult to weld, as specialist equipment is required
copper	malleable ductile good conductor of heat and electricity corrosion-resistant	electric cables plumbing fittings and pipes hot water cylinders	easily drawn into thin wires can be recycled easily soldered	expensive will tarnish (change colour) over time
zinc	excellent resistance to corrosion	protective coverings for railings and dustbins negative battery terminals	can be recycled	brittle
brass (alloy)	good resistance to corrosion good fluidity, casts well good conductor of heat and electricity	plumbing fittings marine fittings	can be polished to achieve a high-lustre finish tougher than copper can be recycled easily cast and turned	relatively expensive

Table 1.5: Properties and uses of non-ferrous metals

Alloys

An alloy is a metal that is formed by mixing two or more metals and sometimes other elements together. An endless list of alloys is possible, each with its own properties such as increased hardness. Alloys are normally grouped as ferrous or non-ferrous. Both mild steel and brass are alloys. This is how they are composed:

Mild steel	99.8% iron 0.2% carbon
Brass	65% copper 35% zinc

The properties of metals

When choosing a material for a specific purpose you must consider its properties. If a company wants a component to be able to withstand impact and sudden shock loading, such as a hammer head, a malleable material would not be suitable. You would need to use a material that is tough, such as high carbon steel.

Figure 1.8: Overhead electricity cables are made from steel-reinforced aluminium because they are ductile

Property	Definition	Materials	Product uses
ductility	the ability to be drawn or stretched into thinner, smaller sections	mild steel copper aluminium	electrical cables
malleability	the ability to be deformed by compression without tearing or cracking	mild steel aluminium	car body panels
hardness	the ability to withstand abrasive wear and indentation	stainless steel high carbon steel	kitchen sinks chisels, hand saws, plane blades, hacksaw blades, scribers and centre punches
toughness	the ability to withstand sudden and shock loading without fracture	mild steel	nails, screws
elasticity	the ability to return to original shape once the deforming force is removed	high carbon steel silver steel	springs

Table 1.6: Property definitions and product uses

Strength	Definition	Materials	Product uses
tension	the ability of a material to withstand being pulled apart	carbon steel	nails, screws, nuts and bolts
compression	the ability of a material to withstand being squashed	mild steel	car body panels
shear	the ability of a material or joint, to withstand being slid or pulled apart	stainless steel	security shear nuts

Table 1.7: The strength definitions of steel and its product uses

Polymers

Objectives

- **Describe** the aesthetics and properties of thermoplastics.

- **Describe** their uses, advantages and disadvantages when manufacturing products.

A polymer is made up from many molecules which are formed into long chains. The way the chains form and the chemical bonds between them lead to differences. Thermoplastics have long, tangled chains of molecules whereas thermosetting plastics have chains that are bonded by short cross-links.

Polymers are generally produced with a glossy, shiny surface, which is impervious or waterproof. In some cases, such as that of polyvinyl chloride (PVC), the addition of plasticisers during manufacture results in a rubbery feel, making it more flexible and ideal for upholstery and clothing. It is possible to produce polymers in an endless range of colours through the addition of pigments during manufacture.

Thermoplastics

Thermoplastics are made up from long chains of molecules that are tangled together and have no fixed pattern. There are very few cross-links between the long chains. This means that when thermoplastics are heated, they become soft, which allows them to be bent, pressed or formed into different shapes. As they cool, they become stiff again. The main advantage of thermoplastics is that they can be reheated and reshaped many times. This feature is one of the key characteristics of thermoplastics.

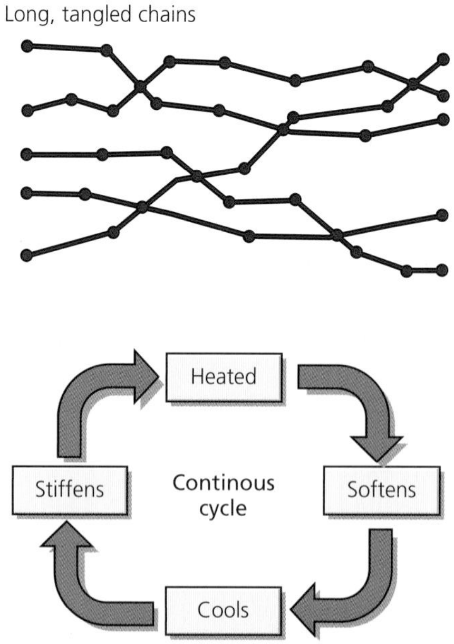

Figure 1.9: A modern house with uPVC windows and fascia boards

Figure 1.10: The structure of thermoplastics

Support Activity

Using the photograph in Fig 1.9 as a start, make a list of all the other applications of PVC that you can think of in your house.

Thermoplastics have a 'memory' and when they are reheated they will try to return to their original flat shape, unless they have been damaged by overheating or overstretching. This property is known as plastic memory.

The properties of thermoplastics

Polymers exhibit a full range of properties depending on their own specific chemical formula. Collectively, they are all good electrical insulators. Specific properties and applications of thermoplastics are shown in Table 1.8.

Polymer	Properties	Uses	Advantages	Disadvantages
acrylic	good impact strength (tends not to shatter but to break into large pieces) lightweight good electrical insulator durable	ornamental fish tanks baths and bathroom furniture car indicator covers/reflectors	can be recycled excellent environmental stability polishes and finishes well available in numerous colours	relatively soft scratches easily poor chemical resistance
polyethene	tough resistant to chemicals soft and flexible good electrical insulator	carrier bags toys washing-up bowls bleach bottles buckets shampoo bottles	can be recycled but not easily excellent chemical resistance	although it can be recycled, most of the waste ends up in landfill sites
polyvinyl chloride (PVC)	good chemical resistance weather-resistant lightweight good electrical insulator stiff hard tough waterproof durable	pipes rainwater pipes and guttering bottles shoe soles window frames and fascias water beds swimming pool toys electrical cable insulation	can be recycled relatively cheap to manufacture	very expensive to recycle dangerous fumes given off when burnt
high-impact polystyrene (HIPS)	tough high impact strength rigid good electrical insulator	food appliances toys cutlery DVD and CD cases	available in numerous colours can be machined and painted can be recycled	expensive limited flexibility will not biodegrade
acrylonitrile-butadiene-styrene (ABS)	high impact strength tough scratch-resistant lightweight durable good resistance to chemicals good electrical insulator	kitchenware camera cases toys car components telephone cases	available in numerous colours	relatively expensive when compared with polystyrene

Table 1.8: The properties and uses of thermoplastics

Thermosetting plastics

These types of plastics are made up from long chains of molecules that are cross-linked. This results in a very rigid molecular structure. Thermosetting plastics will soften when heated, but only for the first time. This allows them to be shaped, but because they are set in a rigid and permanently stiff molecular structure, they cannot be reheated and reshaped like thermoplastics. Two thermosetting plastics are:

- polyester resin
- urea formaldehyde.

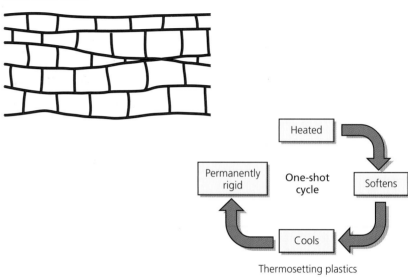

Cross-linked chains

Figure 1.11: The structure of thermosetting plastics

Properties of thermosetting plastics

Polyester resin and urea formaldehyde are both good electrical insulators and have been used to produce plug and plug sockets. They are also both very hard and waterproof when solid, making them ideal for use as adhesives for external applications.

Polymer	Properties	Uses	Advantages	Disadvantages
polyester resin	good electrical insulator hard brittle good heat and chemical resistance resists UV radiation	casting encapsulation for biological specimens boat hulls with fibreglass model figures adhesives filler materials	can be mixed with pigments to achieve a range of colours good resistance to water	contracts on curing can cause excess heat when too much catalyst is used
urea formaldehyde	stiff hard brittle scratch-resistant stain-resistant high tensile strength	tableware worktop laminates buttons electrical casings	can be coloured high surface hardness	toxic fumes given off when it cures

Table 1.9: The properties and uses of polyester resin and urea formaldehyde

Figure 1.12: A cast paperweight made using a clear polyester resin

Choosing a polymer for a specific application requires an understanding of its properties. Inability to change shape may not work for some products. Inability to withstand weathering may shorten the life of some products.

Property	Definition	Materials	Product uses
plasticity	the ability to change shape without cracking or breaking	acrylic ABS	moulded products such as garden chairs, shampoo bottles
durability	the ability to withstand weathering deterioration or corrosion	generally all plastics, although some will fade in colour over time if exposed to UV for too long	garden furniture window and door frames rainwater pipes and guttering

Table 1.10: Property definitions and their product uses

ResultsPlus
Exam Question Report

The handle of a garden spade is made of ABS plastic and is formed using the injection-moulding process.
Explain *one* reason why ABS plastic needs the property of plasticity.
(2 marks, 2007)

How students answered

The overwhelming majority failed to make any comment about plasticity or to give any form of explanation about why ABS needs the property of plasticity.

70% 0 marks

Some comments made about the fact that the plastic needs to be squeezed into the mould but with little or no explanation as to why.

28% 1 mark

Good comment about plasticity and its relevance to being easily injected under pressure into the mould.

2% 2 marks

ResultsPlus
Exam Tip

It is crucial that you learn these definitions and the materials associated with plasticity and durability.

Composites

Objectives

- **Describe** the structural, functional and mechanical properties of carbon fibre and glass reinforced plastics (GRP).

- **Describe** their uses, advantages and disadvantages when manufacturing products.

Carbon fibre

Structure and function

Carbon fibre is a material that is basically made up from very thin fibres, about 0.005–0.010 mm in diameter, and is made mostly from carbon atoms. As many as several thousand fibres can be twisted together to form a thread, which can then be either used on its own, or woven together to form a fabric. Figure 1.13 shows how strands can be woven together to make a mat.

Carbon fibre is a composite material. The matting or strands are combined with a resin known as the matrix, which is most commonly an epoxy resin, to form a new engineered material. This material has new and different physical and chemical properties from those originally shown by the two separate materials.

Properties

Carbon fibre composites produce a material which combines low weight with a high tensile strength. This is known as a high strength-to-weight ratio, meaning that it is very strong when compared with its weight. They can also be formed to create products and components where great stiffness is required, such as in the high performance world of Formula 1 racing cars and in the aircraft industry.

Uses

Carbon fibre has many functions where its properties are used to good effect. Some uses are:

- golf club shafts
- skis
- bike frames, forks and wheels
- yacht and power boat hulls
- racket frames
- fishing rods
- helicopter rotor blades
- aircraft fuselages
- high-quality musical instrument bodies.

Figure 1.13: Woven carbon-fibre matting

Support Activity

Describe why items such as laptops and mobile phones are now manufactured with a carbon-fibre shell or skin.

Stretch Activity

Discuss the benefits for the aircraft industry of using carbon fibre for the fuselage of an aircraft.

Advantages	Disadvantages
• high strength-to-weight ratio	• very expensive
• high tensile strength	• weak when compressed, squashed, or subjected to a high shock or impact
• weave of the cloth can be chosen to maximise strength and stiffness of final component	• small air bubbles or imperfections will cause weak spots and reduce the overall strength
• can be woven in different patterns to create aesthetically pleasing surface patterns	

Table 1.11: Advantages and disadvantages of carbon fibre

Glass reinforced plastic

Structure and function

Glass reinforced plastic (GRP) is another composite material. It is a plastic reinforced with very fine fibres made of glass. Like carbon fibre, the new material is commonly known by the name of the reinforcing fibres, in this case, fibreglass. The plastic which acts as the matrix is most commonly polyester resin or an epoxy resin, both thermosetting plastics.

As with many other composite materials, the two materials – the fibreglass and the resin – work together to overcome the weaknesses of each other.

Properties

Like carbon fibre, GRP is a material which has a good strength-to-weight ratio. The resin is strong in compression and weak in tension. The glass fibres are strong in tension but have little compressive strength. When combined, it forms a new material that resists both tensile and compressive forces. It is a versatile material and, as it can be formed into virtually any three-dimensional shape, it has varied applications.

Uses

With its versatility in terms of shape, fibreglass has endless uses. For example, its lightweight strength and low maintenance requirements make it ideal for architectural mouldings and features which would otherwise be too heavy. Products can also be manufactured in a huge range of colours, and surface textures can be incorporated into the moulds. Uses of GRP include:

- boat hulls
- canoes
- car body panels
- chemical storage tanks
- septic tanks
- train canopies.

Figure 1.13: Composite based products

Advantages	Disadvantages
• lightweight	• difficult to repair
• low maintenance	• time-consuming to make
• endless colours can be achieved	• labour-intensive process
• ability to be formed into almost any 3D shape	• extraction required due to toxic nature of materials involved
• good strength-to-weight ratio	• cutting it produces a fine dust that is dangerous if breathed in
• surface textures can be added to moulds	• requires a mould to be produced to form around
• durable	• resins and catalysts have a limited shelf life
• good resistance to UV light and sea salt	

Table 1.12: Advantages and disadvantages of GRP

Support Activity

Collect a small piece of both woven carbon fibre sheet and chopped strand matting. See how easy it is to pull apart each material and make a few notes on this.

Modern and smart materials

Figure 1.15: SMA orthodontic brace

Shape memory alloys

Shape memory alloys (SMAs) are a collection of metal alloys that can 'remember' their original cold-formed shape. If they become strained or deformed, they can be made to return to their original shape through the application of heat, which will bring about a change in the crystal structure. The heat could be from the human body, an external source or, in some cases, from a small electrical current.

One of the most common SMAs is nitinol, a nickel-titanium based alloy used in the early days for greenhouse window openers.

Uses

Glasses
Nitinol-based glasses frames are claimed to be 'nearly indestructible'. As with most glasses, they are subjected to twisting, bending and sometimes even being sat on. If they have been damaged, they can be gently warmed and will return to their original shape.

Anti-scalding valves
Valves that open and close by using nitinol can be fitted to water taps and shower units. Above a certain temperature the device will automatically shut off the water flow, preventing the user from being scalded by the hot water.

Orthodontic wires
Orthodontic brace wires have traditionally been made of stainless steel. This resulted in visits to the orthodontist every three or four weeks to have the wires retensioned. Now, 'super-elastic' SMAs are being used instead of stainless steel, as you can see in Figure 1.15. As a result of their elastic properties, the SMAs can apply a more gentle pressure over a longer period. This has resulted in less pain for the patient and longer periods between visits to the orthodontist.

Advantages	Disadvantages
• good elasticity • strong in tension • lightweight	• relatively expensive to make in comparison to stainless steel or aluminium

Table 1.13: Advantages and disadvantages of shape memory alloys

Photochromic paint

Photochromic paint changes colour when it is exposed directly to UV light or sunlight. The colour change is reversible – when the UV source is removed, it will go back to its original colour.

Photochromic paint is mainly white, the base colour, and can change colour within one second of being exposed directly to the UV source. The dyes can be added to paints and inks or directly into polymers before being used in the injection-moulding process.

Uses

The paints and dyes are used extensively in the textiles and clothing industry, where T-shirts are printed or painted and will change colour as day turns to night.

Photochromic paints have many other creative and imaginative applications. In Figure 1.16, there appear to be two separate vans, but it is fact the same van. During the day when the UV levels are high, one image appears, but as the sun goes down and darkness falls, a second image appears on the van. The trickery lies in the photochromic paint

Figure 1.16: One van with two paint jobs that change as darkness falls

Advantages	Disadvantages
• change colour in response to UV exposure	• amount of change is dependent upon the level of UV falling on it • over time the ability to change will decay (this is called natural fatigue)

Table 1.14: Advantages and disadvantages of photochromic paints

Quick notes

SMA can change back to their original shape after they have been deformed if the heat is applied. They are flexible yet strong and are lightweight which makes them ideal for making glasses and braces, etc.

Photochromic paint can reversibly change colour in UV light. It is mostly used in textiles companies for clothing design.

Figure 1.17: Photochromic glass lenses that have been partially subjected to UV light

Reactive glass

'Reactive glass' is a term that applies to a collection of types of glass that can change colour in response to exposure to ultraviolet light or to an applied voltage. As with photochromic paint, any change in colour is fully reversible.

The most widespread use of such glass is in spectacles. Reactive glass darkens when exposed to UV radiation. Once the light source has been removed, the lenses slowly return to their original, clear state. This makes it possible for conventional glasses to be used as sunglasses when the user moves outside into a sunny area. You can see the effect of UV on photochromic glass lenses in Figure 1.17, where part of the lens has been exposed to UV light and part has been obscured.

This ability to change is achieved by the addition of silver halide microcrystals to the glass during manufacture.

Another type of reactive glass is changed by the application of a voltage. This type of glass is known as 'smart glass' or 'switchable glass'. Smart glass is used in windows, skylights and offices. At the flick of a switch, the glass changes from being transparent, or see-through, to being opaque, which obscures what is behind the glass. Smart glass can be used to save energy heating costs and, in some cases, can be used instead of curtains and blinds.

Figure 1.18: At the flick of a switch the window glass can be made to change from transparent to opaque

Advantages	Disadvantages
• ability to change colour in response to UV or an applied voltage • replaces the need for separate reading and sunglasses	• expensive to manufacture • smart glass is expensive to install • time delay of photochromic glasses can cause difficulties when driving

Table 1.15: Advantages and disadvantages of reactive glass

Carbon nanotubes as additives to materials

Carbon nanotubes are cylindrical nanostructures made from carbon molecules.

Carbon nanotubes are potentially very useful in the world of electronics, optics and medicine. Due to their mechanical properties, there are proposals to include nanotubes in such items as clothes, sports equipment, and police and military body armour. However, there remains some doubt about their use in medicine due to their potential toxicity.

When nanotubes join together they form long 'ropes'. They can have many forms such as single-wall nanotubes (SWNT) and multi-wall nanotubes (MWNT).

Single-wall nanotubes exhibit electrical properties not shared with multi-wall tubes, and therefore are more likely to be used in continuing the miniaturisation of electrical products. However, they are still very expensive to manufacture but they are becoming cheaper. If it were not possible to reduce the manufacturing cost, it would be very difficult to apply this type of technology on a commercial basis.

Here are some other properties and applications of carbon nanotubes:

- six times lighter than steel, 500 times stronger
- as flexible as plastic
- conduct heat and electricity better than any other material discovered
- can be made from raw materials such as methane gas
- almost totally inert
- used to strengthen plastics on cars
- added to paint to give a very hard, tough finish.

Advantages	Disadvantages
• super tensile strength	• expensive to manufacture
• electrical conductors	• toxic nature may prevent potential applications in the world of medicine
• tough	
• chemically inert	

Table 1.16: Advantages and disadvantages of carbon nanotubes

ResultsPlus
Build Better Answers

Explain **one** advantage for a business of using reactive glass in its office windows. (2 marks)

⚠ **Basic answers (0 marks)**
Misunderstanding the term 'reactive glass'.

● **Good answers (1 mark)**
Suggest that energy can be saved but no evidence or explanation as to why.

■ **Excellent answers (2 marks)**
No need to buy curtains or blinds because the opaque nature of the glass means that people outside cannot see in.

Quick notes
Reactive glass can reversibly change colour in UV light or if a voltage is applied. It is widely used in glasses and windows but is expensive.

Carbon nanotubes are tough and strong, good electrical conductors and chemically inert. They are added to other materials to improve them. They are often used in paint and cars but are expensive and toxic.

exam zone

Know Zone
Chapter 1 Materials and components

Materials and components are fundamental in the manufacture of any product. Great care and consideration must be given to selecting the right material for a given job. Criteria such as the material's aesthetic properties, structure and mechanical properties must all be considered in selecting an appropriate material.

You should know...

the following about the listed materials:

Materials	Properties	Advantages and disadvantages	Uses/applications	Structural composition
Woods	✔	✔	✔	✗
Manufactured boards	✔	✔	✔	✔
Ferrous metals	✔	✔	✔	✗
Non-ferrous metals	✔	✔	✔	✗
Mild steel and brass				✔
Thermoplastics	✔	✔	✔	✗
Thermosetting plastics	✔	✔	✔	✗
Composites	✔ functional, mechanical ✗ aesthetic	✔	✔	✔
Smart materials	✗	✔	✔	✗

Key terms

ductility

malleability

hardness

toughness

elasticity

strength in tension, compression and shear

plasticity

durability

To check your answers, look at the glossary on page 173.

Which of the key terms on the left best fits the descriptions below?

A the ability to withstand abrasive wear and indentation

B the ability of a material or joint to withstand being slid apart

C the ability of a material to withstand being pulled apart

D the ability to withstand sudden and shock loading without fracture

Multiple-choice questions

1. Which of the following is a composite material?

A acrylic

B glass reinforced plastic (GRP)

C mild steel

D copper

2. Which of the following is a thermosetting plastic?

A high impact polystyrene

B polyvinyl chloride

C polyester resin

D polyethene

ResultsPlus
Maximise your marks

The pine body is made in high volume.
The shape of the pine body is suitable for manufacture using CAM.

Describe *one* way in which the shape of the pine body is suitable for manufacture in high volume using CAM. (2 marks, 2009)

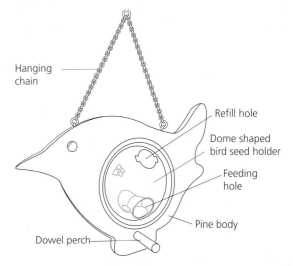

Hanging chain

Refill hole

Dome shaped bird seed holder

Feeding hole

Pine body

Dowel perch

Student answer	Examiner comments	Build a better answer
● *It is flat* (1 mark)	'Flat' is a good start but there is no further comment to describe why this makes it suitable.	▲ The shape is flat and simple which means it can be held easily on the machine.
● *It is a basic and simple shape* (1 mark)	Simple shape is another good reason but no further description.	▲ It is a simple shape which means it can be easily cut.
■ *Pine is easy to cut* (0 marks)	The reference about being easy to cut is not relevant because the question is about the shape of the pine body, not how easy it is to be cut.	▲ The long, curvy shape of the pine body makes it easy to cut because the machine can cut at high speeds.

Overall comment: In many responses, students are often able to give a correct reason but generally fail to go on to fully describe the effect or consequences.

Practice exam questions

1. (a) The soap dish shown below is manufactured from acrylic. Describe **one** reason why acrylic is a suitable material from which to make the soap dish. (2)

(b) Name **two** other thermoplastics that could be used to make the soap dish. (2)

Chapter 2 Tools and equipment
Marking out and measuring

Objectives

- **Recognise** and be able to select the correct tools and equipment when marking out and measuring.

- **Understand** the tools' uses, advantages and disadvantages.

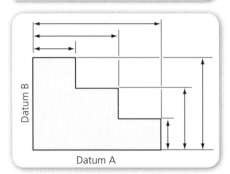

Figure 2.1: Datum edges

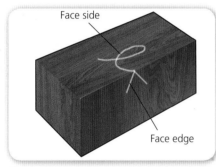

Figure 2.2: Face edges

Marking out and measuring is a critical part of manufacturing and is usually subject to a number of quality control checks. If components are marked out and measured wrongly before being cut out, there is no chance of them fitting together when they are assembled.

Always take marking out measurements from a datum as indicated in Figure 2.1. A datum edge is a flat face or straight edge from which all measurements are taken. This prevents cumulative errors being made.

If you are using timber, choose the face side carefully, before marking it with a small symbol for identification purposes, as shown in Figure 2.2. Then select a face edge that is at right angles to the face side. Take all your measurements from this side and/or edge.

Marking out and measuring tools

These are the tools used for marking out and measuring that you need to know about:

- rules
- squares
- gauges
- scribers.
- punches
- templates
- micrometers

Rules

There are two basic types of rule: steel rule and steel tape. Both start at zero and have millimetre graduations.

Item	Name	Use	Advantages	Disadvantages
	steel rule	for measuring up to 300 mm in length	rigid form which means it will not bend and flex	ends can get worn, so the measurements are not accurate
	measuring tape	for making longer measurements up to 5 m	longer, so more versatile	can become twisted and break ends can break off, making them useless

Table 2.1: The uses of rules

Squares

There are a number of squares:

- try square
- mitre square
- engineer's square.

Both the try square and engineer's square are used to mark lines at 90° to an edge. A try square is used on timber and an engineer's square is used on metals. Both can be used for marking out plastics.

You can also use try squares and engineer's squares to check that a cut or an edge has been made at right angles to another. Hold the stock part of the square tightly against the edge that you have just cut. If you can see light between the two edges then the cut is not square.

A mitre square is used for marking out 45° or 135° angles on wood and plastic.

Take great care when using any form of square for marking out or checking, and ensure that it is being held firmly and tightly against the surfaces or edges of the material.

Stretch Activity

A try square will usually have a brass face on the inside of the stock. Describe why there is a brass face on the stock rather than it being left as wood.

Item	Name and use
	try square marking out or checking right angles on wood or plastic
	engineer's square marking out or checking right angles on metal or plastic
	mitre square marking out or checking angles of 45° or 135°

Table 2.2: The uses of squares

Figure 2.4: A scriber

Gauges

There are three basic types of gauge:

- marking gauge
- mortise gauge
- cutting gauge.

A marking gauge is used for marking lines parallel to the face edge and side on wood. It consists of a stock that slides up and down the stem, allowing various measurements to be set. The gauge should be set using a steel rule that has a zero end. The spur (sharp point) is pushed into the wood as the gauge is pushed or pulled along the length of the timber. It is important to hold the stock tightly against the edge of the timber to ensure that you mark a parallel line.

A cutting gauge is used for cutting across the grain. It is used in the same way as a marking gauge, but has a blade instead of a spur. The blade cuts the fibres across the grain, making it easier and neater to cut with a saw.

A mortise gauge has two pins; one pin is fixed and the other is adjustable. It is used for marking two parallel lines where a mortise and tenon joint is to be cut. The process of marking out is exactly the same as with the two other gauges.

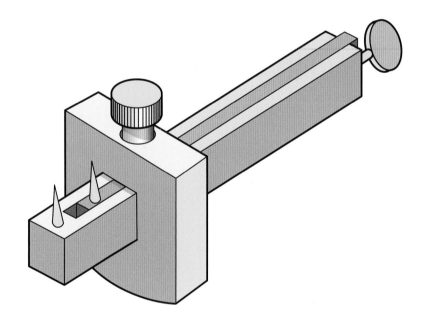

Figure 2.3: Mortise gauge

Scribers

A scriber is used to scratch the surface of metal and plastic lightly. If you are using a scriber on metal, it is a good idea to apply a coat of engineer's blue. This is a spirit-based liquid that is applied to a metal surface. When the scriber is dragged across the engineer's blue it leaves a clean line, which can be easily seen.

Punches

Centre punches are used to make an indent in the surface where holes are to be drilled in metal, as shown in Figure 2.5. They provide a starting point for the drill and stop it skidding over the surface. Dot punches are used for marking the centres where dividers are to be used. They are similar to a centre punch, except that the tips are ground to a 60° rather than a 90° point.

Quick notes
Scribers are used to scratch the surfaces of metal and plastic. Punches are used to indent metal surfaces. Do not mix these up.

37

Figure 2.5: A centre punch, and the punch in use

Templates

A template is used when a number of identical shapes or patterns need to be marked out. You can make a template from any thin material, such as plywood or aluminium, that is easy to draw around.

Micrometers

A micrometer is a specialised instrument used to take very accurate measurements. The thimble, which rotates as the micrometer is tightened, has 50 equal divisions around its diameter, giving an accuracy of 0.01 mm. A reading is taken by adding all the visible whole numbers to the nearest 0.5 mm. The reading from the thimble, which will be between 0 and 0.49 mm, is added to the main reading to get the exact measurement. Although the micrometer provides a very accurate measurement, it can be difficult to learn how to read it.

Figure 2.6: A micrometer

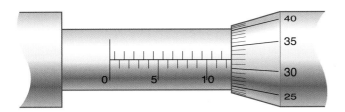

Figure 2.7: The reading on this micrometer is 12 + 0.32 = 12.32 mm

Stretch Activity

Scribers and punches are made from a grade of steel known as 'tool steel' or high carbon steel. Investigate why this is. Which properties of high carbon steel make it suitable to be used in this way?

ResultsPlus
Build Better Answers

Name the tool and describe its use. (2 marks, 2003)

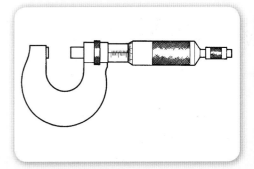

■ **Basic answers (0 marks)**
Simply not able to recall the fact that the tool shown is a micrometer or that the tool is used for measuring or taking accurate measurements.

▲ **Good answers (1–2 marks)**
Tool is correctly identified and named as a micrometer and it is correctly described as being used for measuring or taking accurate measurements.

Wasting

Objectives

- **Recognise** and be able to select the correct tools when removing waste material during the manufacture of products.

- **Understand** the tools' uses, advantages and disadvantages.

ResultsPlus
Build Better Answers

Name the tool and describe its use.
(2 marks)

■ Basic answers (0 marks)
Not able to recall the fact that the tool shown is a tenon saw or that the tool is used for making straight cuts in wood, or cutting tenon joints and shoulders.

▲ Good answers (1–2 marks)
Tool is correctly identified and named as a tenon saw and/or it is correctly described as being used for making straight cuts in wood, or cutting tenon joints and shoulders.

Wasting tools

A wasting process is one that produces waste or unusable material by either cutting bits out or cutting bits off. These are the tools used for wasting that you need to know about:

- saws
- planes
- chisels
- files
- drills
- abrading tools.

Saws

Saws are used to cut material that is not needed away from material which is. Saw blades have alternate teeth bent out or 'set' in opposite directions. This is so that when they cut, they make a gap, called the kerf. The kerf must be wider than the saw blade so that the blade cannot get stuck. When using a saw, you should always cut to the waste side of the marked line so that you leave a small amount for finishing by either sanding or filing. Whatever you are cutting, it is important to keep as many teeth in contact with the piece being cut as possible.

You should choose the correct saw for the type of material you are using. Table 2.3 on the next page shows the most common types of saws used in school workshops.

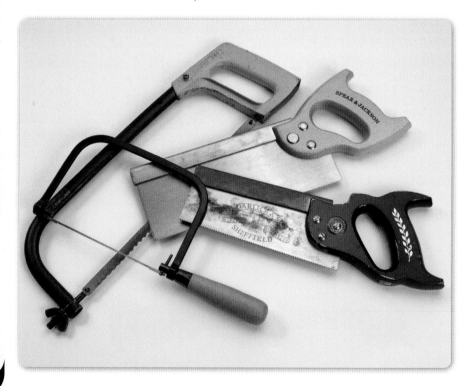

Saw	Name	Use	Advantages	Disadvantages
	coping saw	cut curves in wood and plastics has a thin, replaceable blade held in a frame with the teeth pointing backwards, towards the handle	blade can be rotated easily to cut complex shapes and curves blade can be removed from the frame so that pockets or windows can be cut out	blades easily broken due to their size difficult to control when making straight cuts blades can be put in the wrong way round
	tenon saw	the saw most commonly used for cutting wood in the school workshop 250–350 mm long with 12–14 teeth/mm can be used to cut all general joints; specifically used to cut shoulder of tenon in tenon and mortise joints	good general-purpose woodworking saw	depth of cut limited to depth of the blade
	dovetail saw	used for small, accurate work such as dovetail joints shorter than a tenon saw with 20–25 teeth/mm	fine, accurate cut smaller teeth make it ideal for finer work	only appropriate for fine work not robust enough for general purpose work
	adjustable hacksaw	has replaceable blade held in a frame blade can be angled to cut difficult shapes or if the frame gets in the way of the piece being cut blade can vary in length from 250 mm to 300 mm teeth face forwards and blades have 14–32 teeth/mm can be used for fine work or rough cutting out	can be used to make straight cuts in plastic and metal blade can be removed from frame to cut windows or pockets out	blades can snap or twist easily blades can be put back in the wrong way round

Table 2.3: The most common types of workshop saw

Support Activity

Try cutting a square hole in a piece of plywood. First drill some holes in the wood so that you can pass the coping saw blade through it.

Stretch Activity

Try to work out why it is better to have the teeth of the blade facing backwards towards the handle.

Planes

Planes are used to smooth wood flat and to reduce to size. A jack plane is longer and heavier than a smoothing plane, which is used for finishing and planing end grain because it is easier to handle than a jack plane. A block plane is the smallest plane and is generally used for removing sharp edges and for putting a small bevel along an edge.

Plane	Name	Use	Advantages	Disadvantages
	jack plane	smooths and flattens wood to size	long and heavy so ideal for creating a flat surface	easily jammed if not correctly set heavy
	smoothing plane	finishing a surface and for use on end grain	ideal for use on end grain due to its length	easily jammed if not correctly set
	block plane	removes sharp edges and makes a bevelled edge	can be used in one hand easy to create small bevels	easily jammed if not correctly set not effective on large flat surfaces

Table 2.4: Common workshop planes

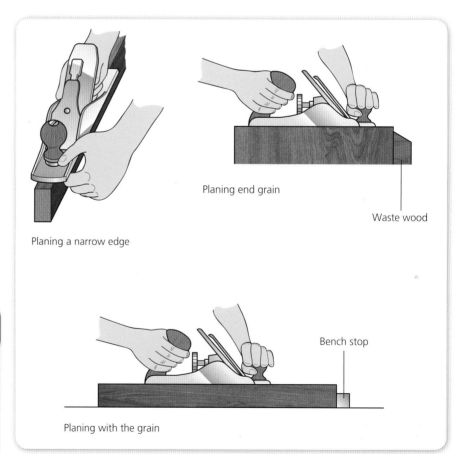

Planing a narrow edge

Planing end grain

Waste wood

Planing with the grain

Bench stop

Figure 2.8: Planing

Chisels

Wood

Four basic wood chisels are used in the school workshop:

- the **firmer chisel** is a general-purpose chisel, which has a square edge
- a **bevel-edge chisel** has a bevelled blade that allows it to get into corners and is especially useful for cutting dovetails
- **mortise chisels** have much deeper blades and are used with a mallet for cutting mortise joints
- **gouges** have curved blades and are used for carving.

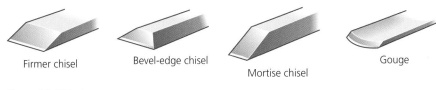

Firmer chisel   Bevel-edge chisel   Mortise chisel   Gouge

Figure 2.9: Chisels

Metal

Cold chisels can be used to cut sheet metal, either by shearing across it or by chopping down on it vertically. They have a hardened and tempered cutting edge while the other end is left soft to absorb the impact from the hammer blows. Different profiles are available, allowing access to corners or for producing grooves in the workpiece.

Abrading tools

Abrading tools remove very small particles of waste, such as those produced by filing. Abrading tools include rasps, which are used on wood, and surforms. Surforms have replaceable blades. They are formed and operate a bit like a cheese grater.

Abrasive papers are also available for wood, metal and plastics. Glasspaper, sometimes called sandpaper, is used on wood and is made from glass particles stuck to a thick paper sheet. The paper is graded according to how much grit there is per unit area. Emery cloth is used on metals and plastics and it is graded in the same way as glasspaper. Any form of abrasive paper is best used wrapped around a cork block. This ensures that an even pressure is applied over the work.

41

Abrading tool	Name	Use	Advantages	Disadvantages
	rasp	quick removal of waste wood	can be used like a file to create external curves good on flat edges and surfaces	clogs up easily
	surform	fast removal of soft material	can be used on flat or 3D surfaces	blades can break easily

Table 2.5: Common workshop abrading tools

Files

Two basic filing processes are used in the workshop:

- cross-filing
- draw-filing.

Cross-filing removes waste rapidly. You should use the whole length of the file with a downwards force. The file only cuts forwards. It should be lifted off at the end of the stroke and not dragged back across the workpiece.

Draw-filing removes marks in the work left as a result of cross-filing. This method gives a much better surface finish, and a smoother file should always be used for draw-filing. An even finer finish can be obtained by wrapping a piece of emery cloth or wet and dry paper around the file and repeating the action.

Files are made from high carbon steel. The main body is hardened and tempered and has rows of teeth. The tang of the file is left soft and fits into the handle.

General work is carried out with a flat file. One of the long edges has teeth; the other is plain and is known as a safe edge. The safe edge prevents cutting into the face of a square corner. Files with different profiles are available for a range of applications. There are also other, more specialist forms of file.

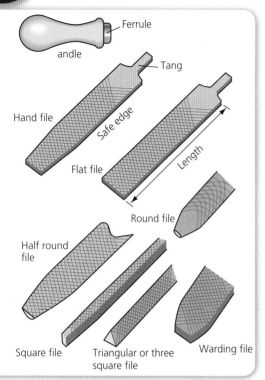

Figure 2.10: Files

File	Name	Use	Advantages	Disadvantages
	flat file	removes waste on large, flat surfaces quickly can be used to create external curves	has a safe edge on one side to prevent cutting into edge when filing a corner	brittle and easily broken if dropped teeth clog when filing soft material such as brass and aluminium (called pinning)
	round file	creates curves and fillets	increases in diameter along the length so can be used on different-sized holes	small cross-section makes it weak and easily broken
	three square file	cuts into corners less than 90°	useful between angles of 60° and 90°	cannot file angles smaller than 60°

Table 2.6: Common workshop files

Drills

Twist drills, the most common type of drill used in school workshops, can create holes in most materials.

There are many other forms of drill, each with its own specific use and application. The most common types used in school workshops are shown below.

Drill	Name	Use	Advantages	Disadvantages
	twist drill	twist or flute carries waste material away in the form of swarf smaller sizes can be used in a hand drill	available in sizes from 0.5mm diameter to over 25mm	small drills prone to breaking if not used correctly
	flat bit	used to drill a hole all the way through, but will break through, splintering wood on underside	fast removal of waste with an electric drill	will split wood on underside when drill breaks through if not supported by another piece of wood
	countersink bit	creates depression for head of a countersunk screw so that it sits flush with surface of material	can be used to create range of countersunk hole sizes	will chatter if used at too high a speed
	hole saw	used on thin materials or to make large hole up to 150mm in diameter used in electric drill	widely used by plumbers and electricians to make holes for pipes and ducting can be used on thick and thin sections of wood and plastic	have tendency to burn if used at too high a speed only come in set sizes

Table 2.7: Common workshop drills and bits

ResultsPlus
Exam Question Report

Name the tool shown here.
(1 mark, 2007)

How students answered
The overwhelming majority of students scored very well on this question.
Drill, drill bit and twist drill were all good answers for 1 mark.

85% got 1 mark

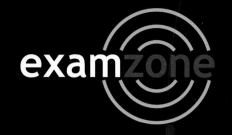

44

You must use tools and equipment correctly when carrying out any manufacturing processes in the workshop. Using the correct tool for the job will help you to measure and mark out successfully and accurately. Careful tool selection when carrying out a wasting process will help you to remove material safely and appropriately.

You should know...

the following about the listed marking and measuring tools and items:

Marking out and measuring tool	Appearance	Uses	Processes	Advantages and disadvantages
Rules	✔	✔	✔	✔
Squares	✔	✔	✔	✔
Gauges	✔	✔	✔	✔
Scribers	✔	✔	✔	✔
Punches	✔	✔	✔	✔
Templates	✔	✔	✔	✔
Micrometers	✔	✔	✔	✔

the following about the listed wasting tools:

Wasting tool	Appearance	Uses	Processes	Advantages and disadvantages
Saws	✔	✔	✔	✘
Planes	✔	✔	✔	✘
Chisels	✔	✔	✔	✘
Files	✔	✔	✔	✘
Drills	✔	✔	✔	✘
Abrading tools	✔	✔	✔	✘

Multiple-choice questions

1. Which of the following is *not* a type of gauge?

A marking   **C** mortise

B measuring   **D** cutting

2. Which type of saw is shown below?

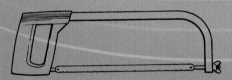

A coping saw   **C** jigsaw

B dovetail saw   **D** adjustable hacksaw

ResultsPlus
Maximise your marks

A student needs to mark out 10 identical copies of the shape.

The shape is marked out onto an acrylic sheet using a template and a marking-out tool.

(i) Name *one* marking-out tool that can be used to mark around the template. **(1 mark, 2007)**

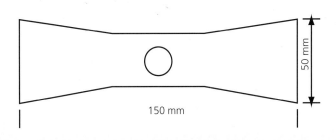

150 mm

50 mm

Student answer	Examiner comments	Build a better answer
■ *Ruler* (0 marks)	This shows a basic lack of understanding of what the question is asking for, and the student has not grasped the concept that a template is being used and it needs to be marked around.	▲ A straightforward response is required, naming the correct tool.
▲ *Scriber* (1 mark)	The scriber is the most appropriate tool to be used since its sharp point will make a visible scratch in the surface of the acrylic.	▲ The most common response seen.

Overall comment: This is a very straightforward question. It relies on your ability to learn and recall factual knowledge. Here you are being asked to apply your knowledge and understanding to an unfamiliar situation. It was very surprising to see that only 30% of students scored the 1 mark available for this question.

Practice exam questions

1. Complete the table below which shows some tools by filling in the missing names and uses.

Tool	Name	Use
	tape	
	plane	
		marking out two parallel lines on wood
		making very accurate measurements

Chapter 3 Industrial and commercial processes
Scale of production

Objectives

- **Recognise** the characteristics of the three scales of production.

- **Understand** their uses, advantages and disadvantages.

Scale of production is very important when designing a new product, as it can influence many features of the design. The level of demand for the product affects both the design and the manufacturing process.

Three scales of production

There are three scales of production:

- one-off
- batch
- mass.

One-off

One-off production, as its name suggests, is the design and manufacture of a single item or product.

Sometimes a client commissions a designer to produce a bespoke or one-off item for a special occasion. Sometimes designers create a one-off piece for an exhibition, either to sell or to use to attract new clients. Sometimes a piece might be a starting point in a discussion for a commission. Figure 3.1 shows an example of an item specially made for an exhibition, a hand-woven silver choker.

Figure 3.1: A one-off handmade silver choker

This piece is made from two different-sized fine soft sterling silver wires. The type of silver used will tarnish less than normal silver – an important design factor, given the intricate nature of the piece. The outside wire frame was made to a template in a thicker wire before being silver-soldered at the corner. The hand-woven fine-wire mesh was then sewn into the wire frame before the freshwater pearls were stitched on. The clasp at the back is a commercially available product and was sewn onto the frame. This clasp allows the choker to be adjusted to fit different-sized necks. This single piece took seven hours to make in total.

This product is labour-intensive, which means that it will be expensive to buy, but there will be no other one like it. Although it requires a high level of skill, it requires a relatively low level of investment in terms of specialised equipment and machinery.

Batch

Batch production is a system of manufacturing used to produce a fixed quantity or 'batch' of identical products. A batch may be small or large depending on the product being produced.

Roof trusses for the construction industry are an example of a batch-produced item. Not all new houses have the same roof size, pitch or shape, so specialist manufacturers make different sizes in batches to meet the architects' requirements. These batches are made to meet builders' schedules; the manufacturer does not make all the trusses for the development at once. The workforce that makes the trusses is probably highly skilled. The manufacturer may also use the same tools and machinery to make other products.

Figure 3.2: Roof trusses for a house

Batch production fits in between one-off and mass production, although a batch may run into several thousand products or components. The economics behind batch sizes are quite complicated but can be explained in relation to manufacturing costs. There is a point, known as the 'break-even' point, below which it is not viable to produce a given quantity. This point is based on set-up costs, which include tooling. Below the break-even point, manufacturing costs are too expensive. Above it, the costs of manufacturing decrease, which makes the product more affordable in the retail market. Careful market research that is carried out before making a decision about a product's market potential will give some indication of how many products are likely to be sold.

Batch production allows companies better control over their cash flow. They do not have to invest lots of money in making products that are then stored for a long time in shops or warehouses. Companies do not have to invest too much money in stock if they watch stock levels and control stock carefully. Many companies now use ICT to control stock levels and order new stock automatically.

In batch production, tools and machines are usually set up specifically to produce the parts and components needed for a particular product. They are then used to produce parts for other products. Tools that are specific to the particular product will be stored away for later use should another batch of similar components or products be required.

Not all machines and manufacturing processes are suited to batch production. Paying people is expensive, so labour-intensive processes mean higher costs. Some products can be machined directly using CNC machinery such as milling machines and lathes. This means that the saved file can be uploaded and the required number machined.

Build Better Answers

Explain one reason why manufacturers make products in batches. (2 marks)

■ **Basic answers (0 marks)**
No reason or benefit given as to why products are made in batches.

● **Good answers (1 mark)**
Offer an idea about why products are made in batches but offer no explanation as to the benefits.

▲ **Excellent answers (2 marks)**
Ideas suggested are correct, including issues related to costs and savings, which are fully explained.

Figure 3.3: A production line (assembly line) where separate items are made into working products

Mass

Mass production (or high-volume production) means making the same product continuously.

Some products made in this way, such as fizzy drink bottles or plastic washing-up bowls, require little or no further assembly or finishing.

Other products, such as television remote controls or mobile phones, have to be assembled by hand along a line. This process is often called a production line, or assembly line (see Figure 3.3).

Production lines sometimes run continuously, 24 hours a day. This is called continuous or flow production. Car manufacturers often run continuous production lines, and their workers have to work shifts.

Making products or components in large volumes often means using very expensive machinery, tools and moulds. These high initial 'tooling up' costs, which cover the cost of the machine itself and any moulds that are used, have to be recovered in the overall cost of the product. However, the actual manufacturing and material costs are generally low.

Some companies, such as mobile phone or motor manufacturers, do not usually make all the components for their products. Car manufacturers, for example, often buy in a large number of the components they need – such as tyres from a specialist tyre manufacturer – but they are still responsible for putting the product together.

Production scale	Characteristics	Advantages	Disadvantages
one-off	highly skilled workers constant discussion with client specialist area of work such as furniture, jewellery	unique product each time products can be made to measure	very expensive you might have to wait a long time for the item to be made
batch	production-line set-up workers semi-skilled and flexible parts often bought from other companies and assembled	quick response to customers' demands production line can be changed quickly	tools need to be reset for new production run
mass	semi-automated with much computer control high level of investment in machinery	low unit costs can operate 24/7	initial machinery and tooling costs are very high

Table 3.1: Characteristics, advantages and disadvantages of different scales of production

Support Activity

Take a product such as a shampoo bottle and describe two features that indicate that it has been mass-produced.

Stretch Activity

Take an item such as an electrical three-pin plug. Take it to pieces and record all its separate parts. Try to work out which parts might have been made by the plug manufacturer and which might have been bought in as special components or items.

Materials processing and forming

Materials processing and forming

Objectives

- **Know** how and when to use various processing methods and the preparations involved.

- **Describe** their characteristics, advantages and disadvantages.

Stretch Activity

Produce a flowchart to show the process of casting. You could look on the internet for some movie clips to guide you.

ResultsPlus
Build Better Answers

Explain **one** reason why it is essential that a high-quality finish is achieved on the mould before casting. (2 marks)

■ **Basic answers (0 marks)**
No understanding at all of the process of casting or any of the implications of having a high-quality surface finish on the mould.

● **Good answers (1 mark)**
An idea is suggested as to why a high-quality surface finish is required but no further detail is given to explain why it is essential.

▲ **Excellent answers (2 marks)**
An idea is given such as that any defects appearing in the mould will also appear in the cast product, and this is fully explained to say that the product is likely to be rejected, wasting time and effort.

Processing and forming methods can be divided into two main areas: deforming and reforming.

- **Deforming** processes such as laminating and vacuum forming allow the material's shape to be changed without changing its state.
- **Reforming** processes such as casting involve a change of state within the material being used; in other words, the material changes from a solid to a liquid or plasticised state.

Processing and forming methods

You need to know about the following methods:

- sand casting
- drilling
- turning (wood and metal)
- blow moulding
- injection moulding
- vacuum forming
- extrusion (plastic and metal)
- wood laminating.

Sand casting

Casting is nearly always used when producing complex shapes such as car engine parts and kitchen and bathroom taps.

Casting aluminium in sand moulds is a relatively cheap and simple process that can be carried out in school workshops. However, you need to make a pattern of the final product first, and this can be difficult and time-consuming.

Advantages	Disadvantages
• any waste can be reused • complex shapes and forms can be achieved • hollow products can be achieved with the use of cores	• consumes a great deal of energy in order to melt metal • new sand cast has to be packed each time the product has to be made • time-consuming to pack sand each time • secondary machining often required to produce flat surfaces

Table 3.2: Advantages and disadvantages of casting

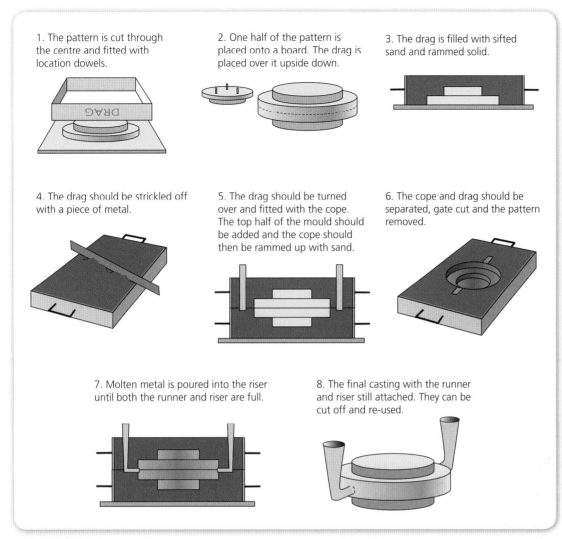

1. The pattern is cut through the centre and fitted with location dowels.

2. One half of the pattern is placed onto a board. The drag is placed over it upside down.

3. The drag is filled with sifted sand and rammed solid.

4. The drag should be strickled off with a piece of metal.

5. The drag should be turned over and fitted with the cope. The top half of the mould should be added and the cope should then be rammed up with sand.

6. The cope and drag should be separated, gate cut and the pattern removed.

7. Molten metal is poured into the riser until both the runner and riser are full.

8. The final casting with the runner and riser still attached. They can be cut off and re-used.

Figure 3.4: Casting

The process can be broken down into the following stages:

- make a pattern of the required product
- encase the pattern in moulding sand
- split the sand box and remove the pattern to leave an empty cavity
- pour molten metal into the mould
- when the metal has solidified and cooled, remove the product.

The quality of the final product depends on the quality of the mould/ pattern used (the same applies to vacuum forming). Patterns can be either single-piece or split. If you are making a split pattern, insert some dowels to act as location points when lining up the mould. Make a draft angle on any pattern because this helps when removing it from the sand. Round off external corners and put a fillet on internal corners to avoid sharp edges. The pattern is often painted or varnished to produce a smooth surface finish.

Drilling

Drilling is one of the processes you are most likely to use in the school workshop. Hand drills, cordless power drills and the pillar drilling machine can all be used in combination with relevant drill bits.

The pillar drill, as its name suggests, stands on a central pillar. A machine vice, supported on a table, holds the object being drilled. A motor, connected to the chuck via pulleys, provides the power. Using a pillar drilling machine can be dangerous and you must take a number of precautions. Some of these precautions will also apply to using a hand drill and a cordless drill, such as wearing goggles and tucking in any loose clothing.

● Use a chuck key to tighten and remove the drill bits from the chuck, but never leave it in the machine.
● Always wear protective equipment such as goggles and an apron.
● Tuck loose items of clothing such as ties out of the way and fasten long hair back.
● Make sure that the object being drilled is held tightly and securely in the machine vice.
● Use the correct drill speed for the material being drilled and for the size of drill being used.

Apply it!

In Unit 1: Make activity you may need to use a pillar drill. As a general rule, the bigger the drill bit, the slower the speed should be.

ResultsPlus
Build Better Answers

Name the tool shown below. (1 mark)

■ **Basic answers (0 marks)**
No mark awarded for just 'drill' since it is a drill *bit* or piece that goes into this machine.

▲ **Excellent answer (1 mark)**
'Pillar drill machine' is the correct name but credit would also be given for 'pillar drilling machine'.

Drill	Name	Advantages	Disadvantages
	pillar drill	can be used continuously chuck can be removed to accept larger drill bits speed can be varied	fixed in one position can be dangerous
	hand drill	portable	you need to provide the power will only hold bits up to 10mm
	cordless drill	portable generally has a variable speed control can be made to reverse direction can be used to put in and take out screws	power limited by battery type and size charging batteries can take some time will only hold bits up to 10mm

Table 3.3: Advantages and disadvantages of workshop drills

Turning

Wood turning

Wood turning allows circular wooden products to be made, such as fruit bowls and stair spindles. There are two different ways to turn wood on a lathe – on a faceplate and between centres – but you cannot use them both at the same time.

The 'outside' spindle is designed to take a faceplate. You can fit a wooden blank onto this to make items such as fruit bowls, vacuum-forming moulds or fibreglass moulds. Figure 3.5 shows a fruit bowl being turned on a faceplate. Great care has to be taken to ensure the safety of the user, who must wear a protective face visor. You must do this to make sure that pieces of wood do not get into your eyes.

Figure 3.5: A fruit bowl being turned on a faceplate

You must take great care when carrying out any form of turning, especially wood turning. The distance between the object being turned and the tool rest must be kept to a minimum to ensure that there is no chance of the tool catching or snagging and being pulled out of the operator's hands. It is also important to make sure that the tool rest and tool are fixed at the correct height.

Turning is wasteful and the waste that is removed is not very useful.

Turning between centres allows long pieces of work to be supported at both ends. A fork provides the driving motion to rotate the workpiece. The other end is held by a 'dead' centre. Stair spindles and lamp centres are turned in this way. Items such as these, where they are all the same, are often made on computer numerically controlled (CNC) lathes, where a computer automatically controls all the tool movements.

Preparation is essential when turning between centres. Prepare the wood so that it fixes onto the lathe without slipping. To do this, make saw cuts and remove the corners by planing, as shown in Figure 3.6.

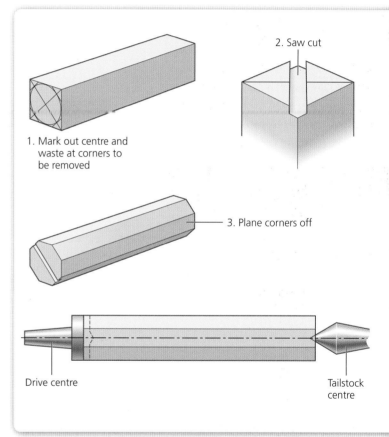

1. Mark out centre and waste at corners to be removed

2. Saw cut

3. Plane corners off

Drive centre

Tailstock centre

Figure 3.6: Preparation for turning between centre

Advantages	Disadvantages
• waste can be reused for other purposes such as animal bedding • one-off shapes and products can be produced • small off-cuts from the workshop can be turned into other products	• long objects can flex if not correctly supported along their length • once an object has been removed, it is difficult to reset it to centre • if correct speeds and tool feed rates are not used, work and tools can be damaged • tools need to be sharpened regularly • workpiece and tools can get very hot • difficult to make exact copies by hand • tools can catch on edges, which can be dangerous

Table 3.4: Advantages and disadvantages of wood turning

Metal turning

When turning metal, a centre lathe is set up in the same way as a wood lathe except that it has no outside spindle. Work can be turned between centres or held in a chuck connected directly to the motor and gearbox. Although you can turn any shape of work on a centre lathe, work in school workshops is usually cylindrical. This means that a three-jaw self-centring chuck is ideal. The three separate jaws all move together, ensuring that the centre line remains constant.

You should use an independent four-jaw chuck to turn square sections, and a faceplate for irregular-shaped items. Both these methods take a long time to set up.

The two turning processes most often used in school workshops are:

- facing
- parallel turning.

Other turning processes include:

- taper turning
- drilling
- parting
- knurling.

Operation		Description
facing		tool moved at right angles to axis, facing end surface
parallel turning		tool moved parallel to axis to form a cylinder, reducing the diameter
taper turning		tool moved at angle to axis to produce a taper
parting		narrow tool fed into work to trim to length or part work from stock bar
drilling		tail stock used to hold chuck into which a drill can be placed; as work rotates, drill is fed into the work
knurling		hardened steel wheel pressed into the rotating work to produce straight or diamond pattern

Table 3.5: Basic operations that can be carried out on a centre lathe

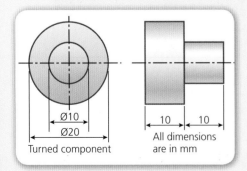

Ø10
Ø20
Turned component

10   10
All dimensions are in mm

Figure 3.7: Turned component

Advantages	Disadvantages
• chuck can hold a variety of sizes and lengths	• very difficult to set up non-cylindrical items
• very accurate sizing can be achieved with fine adjustments of the cross slide	• long objects can flex if not correctly supported along their length
• drilling and tapping can be carried out	• once an object has been removed, it is difficult to reset it to centre
• holes will be drilled in line with the axis of the work	• if correct speeds and tool feeds are not used, work and tools can become damaged
• cylindrical cast items can be turned to achieve a smooth surface finish	• tools need to be sharpened regularly
• bigger tubes can be held using the outside of the three-jaw chuck	• workpiece and tools can get very hot

Table 3.6: Advantages and disadvantages of metal turning

Stretch Activity

Having produced the flowchart for the component shown on page 54, try making it on the centre lathe.

ResultsPlus
Exam Question Report

The table below shows *two* of the turning processes used to manufacture the handle of the screwdriver. Complete the table by naming the correct process given by each description. (2 marks, 2007)

Process description		Process name
[diagram]	The tool is moved at right angles to the centre across the end of the bar.	....
[diagram]	The tool is moved along to reduce the diameter of the bar.	....

How students answered

First process
This type of question requires a basic level of turning and the majority of students were unable to give the correct answer.

████████████████████	90%	0 marks

A very simple and basic response given.

▪	10%	1 mark

Second process
Here the question is trying to find out whether students know more than one turning process, so fewer students gave the correct answer having scored 1 mark on the first process.

█████████████████████	96%	0 marks

Fewer students scored a mark here but again only a basic level of knowledge of turning was required.

▪	4%	1 mark

Support Activity

See if you can find out how many fizzy drink bottles are produced (a) in the UK and (b) in the world every year.

Stretch Activity

Find out what happens to all these bottles when they are empty.

Blow moulding

Blow moulding is a manufacturing process used to make strong, hollow plastic products such as bleach, disinfectant and shampoo bottles and water butts.

The blow moulding process starts with the production of a parison. The parison is a tube-like piece of plastic, a bit like a hosepipe, with a hole in one end into which compressed air can pass. The mould traps the parison before air is blown in to form the shape.

Fizzy drink bottles are blow-moulded following the sequence outlined below.

- The parison is extruded from a plastic injection mould, which is placed very close to the blow-moulding machinery.
- The parison is then lowered into the open mould.
- The two halves of a bottle-shaped metal mould close around the parison, cutting it to length at the same time, as a hollow ramrod is forced into its centre.
- Compressed air is forced out through the hollow ramrod.
- The plastic is forced out to the sides of the mould. As the stretching takes place evenly, the plastic remains an even thickness.
- The bottle takes the shape of the mould and is then dropped out of the blow-moulding machine as the two mould halves open.
- Meanwhile, a new parison is being extruded and the entire process begins again.

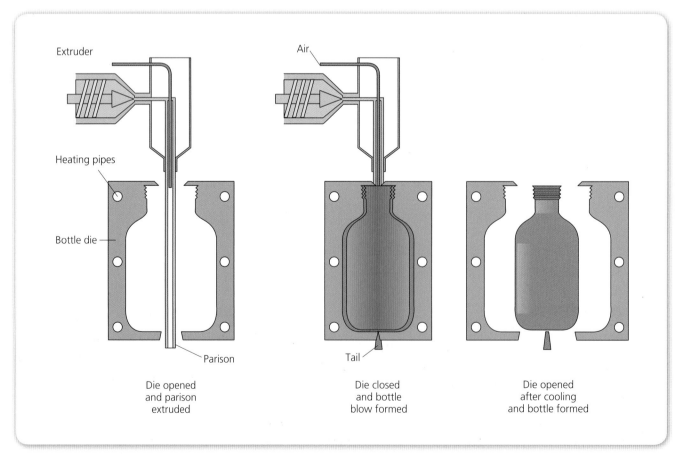

Figure 3.8: Stages of blow moulding

Advantages	Disadvantages
• very cheap unit costs • highly automated process • not very labour-intensive • ideal for high volume, continuous production since it can run 24 hours a day • very little secondary finishing required due to nature of plastic material and mould surface finish	• initial costs of machine and tooling very high • not suitable for small production runs • sometimes a seam is left around product where mould closes • products sometimes need to have flashing removed

Table 3.7: Advantages and disadvantages of blow moulding

There is another, simpler form of blow moulding, which is quite similar to the vacuum-forming process. This involves blowing compressed air through a small inlet valve onto softened plastic. There are several variations of this technique, but all involve forming a dome.

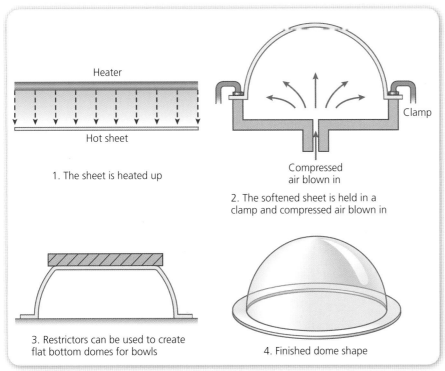

Figure 3.9: The blow-moulding process

As you can see from Figure 3.9, a thermoplastic sheet is heated and then clamped tight and flat. Compressed air is blown in from underneath to form a dome. The volume of air blown in controls the shape and height of the dome, but the air pressure must be constant once the desired shape has been achieved so that it holds its shape. In some instances a form of restrictor is placed over the sheet to restrict the height and shape of the dome. In this example, bowl shapes are being formed.

In commercial applications, the sheet is more often fixed over a hollow mould and the compressed air blown onto the top of the heated sheet, stretching it into the mould.

Support Activity

Collect a range of hollow plastic products. Examine the types of features that have been moulded into them, such as surface texture, Braille and recycling information.

Exam Question Report

Explain *two* advantages of producing the handle using the injection-moulding process. (4 marks, June 2009)

How students answered for the first advantage

Most students answered this question poorly. They often gave no valid response.

| 31% | 0 marks |

Many students were able to give a basic advantage that related to items being identical or accurate.

| 60% | 1–2 marks |

Very few students were able to explain an advantage fully.

| 9% | 3–4 marks |

Injection moulding

This highly automated process is used to produce a wide range of common items such as washing-up bowls, buckets and cases for household electrical goods. Injection moulding is best suited to thermoplastics, but some thermosetting plastics can be used.

The injection moulding machine is made up of:

- a hopper unit
- a screw and injector unit
- a heating element
- a mould.

The process is quite simple. It is repeated continuously to produce moulded products of a high quality that require no further finishing other than removing any sprue pins.

- A hopper full of plastic granules is used to feed a rotating screw mechanism. As the plastic is moved along the screw thread, it becomes plasticised by the heat from the heating element.
- The rotating screw forces the plastic forward into an area ready to be injected into the mould. The screw mechanism also acts as a ram, and this injects the plasticised material into the empty mould.
- The mould is left to cool; sometimes this is achieved by pumping water through parts of it.
- When the temperature is low enough, the mould splits open and the moulded product is ejected by ejector pins.
- The mould then closes and the whole process is repeated.

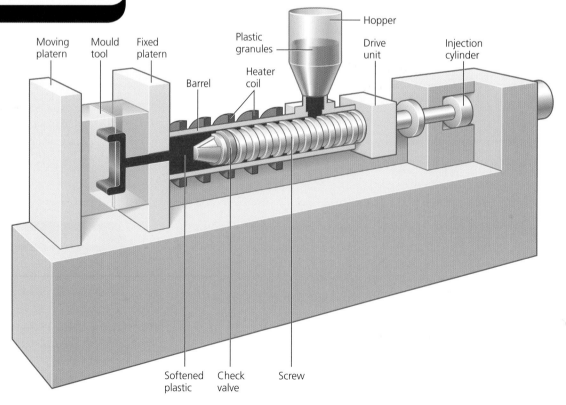

Figure 3.10: Injection moulding

Injection moulding is a versatile process that can be used to produce all kinds of items from pencil sharpeners to industrial-sized wheelie bins.

The single-piece garden patio chair shown in Figure 3.11 is produced in high volume by injection moulding. The mould required is a 3D mould, which only consists of two parts. The chair is a single-piece moulding that might take up to 60 seconds to produce. Various colour pigments can be added to the polymer pellets before they are melted; for example, these garden chairs are moulded in both white and green.

Advantages	Disadvantages
• can operate 24 hours a day • can be used to make different-coloured products • inserts, such as screwdriver blades, can be moulded directly into handles • several smaller items can be manufactured in a single mould • suitable for high-volume continuous production • high level of accuracy • identical components formed each time • little or no secondary surface finishing is required • unit costs low in comparison with initial set-up costs	• initial machine and mould costs high • some flashing may have to be removed • sprue pins need to be cut off

Table 3.8: Advantages and disadvantages of injection moulding

Vacuum forming

The vacuum forming process is used to make various packaging items such as Easter-egg containers, yoghurt pots, trays, dishes and masks. Generally, the most suitable materials for vacuum forming are thermoplastics such as polyethene, PVC, ABS and acrylic. Vacuum forming, pictured in Figure 3.13, is a simple workshop process and follows the basic steps shown below.

a) The mould is placed on the table or platen and lowered down. A thermoplastic sheet is clamped down and held tight before heaters soften it.

b) As the sheet becomes soft and pliable, compressed air is blown up to stretch the plastic evenly before the table is raised, pushing the mould into the sheet.

c) A vacuum pump then removes the trapped air, which in turn reduces the pressure below the clamped sheet.

d) The higher atmospheric pressure above forces plastic down over the mould, forming the required shape.

Figure 3.11: Example of an injection-moulded object

59

Quick notes

Injection moulding:

• is a highly automated process

• can run 24/7

• produces very accurate, identical components

• is expensive to set up but relatively cheap to run.

Apply it!

If you use vacuum-forming in your Unit 1: Make activity make sure you include some photographs and details of the stages involved in making your mould.

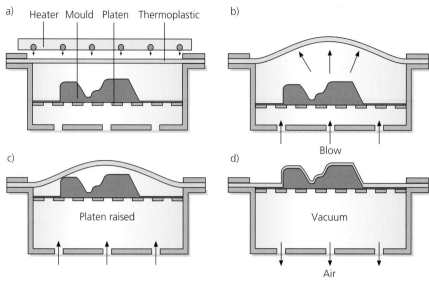

Figure 3.12: Vacuum forming

Support Activity

Figure 3.13 shows a vacuum-formed food package. Draw a cross-section of the mould that would have been used to form the package. Remember to make sure that it incorporates all the design features identified above.

Figure 3.13: Vacuum-formed food package

The quality of mould design and surface finish will determine the success of the final product. Sticking to some basic design principles will result in a more successful product.

- All vertical surfaces need to be slightly tapered (about 5°). This is called a draft angle and makes it easier to remove the mould once the product has been formed.
- Round off sharp corners so that they do not puncture or rip the plastic.
- Incorporate vent holes to avoid pockets of air becoming trapped, which would stop the plastic forming correctly over the mould.
- Round off internal edges and avoid sharp corners in order to help the plastic form over the mould surfaces.

All these points are highlighted in the cross-sectioned mould shown in Figure 3.14.

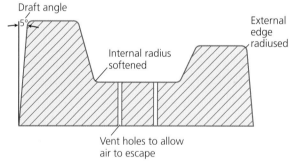

Figure 3.14: Cross-section of a good mould for vacuum forming

Advantages	Disadvantages
• lightweight, hollow products can be made • relatively cheap moulds can be made from MDF in the school workshop for a one-off item • surface textures can be moulded into products	• thermoplastic material sometimes thins too much and may burst or pop • webs sometimes form between items, meaning that the formed product cannot be used • products need to be trimmed and cut out

Table 3.9: Advantages and disadvantages of vacuum forming

Wood laminating

Wood laminating involves building up thin layers around a former to produce the desired shape or curve. Thin veneers, or skin ply, are cut to the required shape, making sure that the grain is running in the same direction, following the curve.

These layers are glued together with an adhesive such as PVA or Cascamite. They are then trapped in a former or jig and held under pressure by clamps. The whole item is left while the adhesive sets. For larger objects, a vacuum table or bag can be used. The layers are prepared in the same way and placed over the former. The air is sucked from the bag using a pump. Atmospheric pressure forces the layers together and around the former while the adhesive sets.

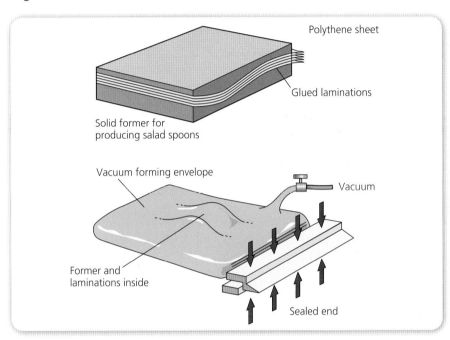

Polythene sheet

Glued laminations

Solid former for producing salad spoons

Vacuum forming envelope

Vacuum

Former and laminations inside

Sealed end

Figure 3.15: Laminating

Laminated timber is also used to create large sectional beams, for example 300 × 255 mm. Timber laminated in this way is used for structural purposes such as roof beams, floor beams or joists. Using laminated timber means that exact dimensions can be achieved either to a specific set of measurements calculated by a structural engineer or as a replacement to match an original part. Large laminated sections can be curved while they are being made, where they are used for both structural and aesthetic purposes.

Advantages	Disadvantages
• complex shapes can be achieved	• requires a former to be made to form over, even for a one-off
• large sections can be built up to improve mechanical strength	• special adhesives have to be used if components are being put outside
• several small products can be laminated at the same time	• must be left for up to 24 hours for adhesives to set

Table 3.11: Advantages and disadvantages of laminating

ResultsPlus
Build Better Answers

Salad servers are laminated from thin pieces of plywood. Choose **one** adhesive from the list below which is best suited to gluing the pieces of plywood together. (1 mark)
• Tensol® cement
• PVA
• contact adhesive
• epoxy resin

■ **Basic answers (0 marks)**
Incorrect adhesive selected.

▲ **Good answers (1 mark)**
PVA is the correct choice.

Figure 3.16: Laminated timber used for flooring

Extrusion

Extrusion is a process used to manufacture products that have a regular, fixed cross-sectional profile such as rain water guttering and copper pipes. The material can be pushed or drawn through a die of the required cross-section. Extrusion can produce continuous lengths of product, or they can be cropped or cut to length. The process can be carried out hot or cold, depending on the material being used. The most common extruded materials are metals, polymers, concrete and foodstuffs.

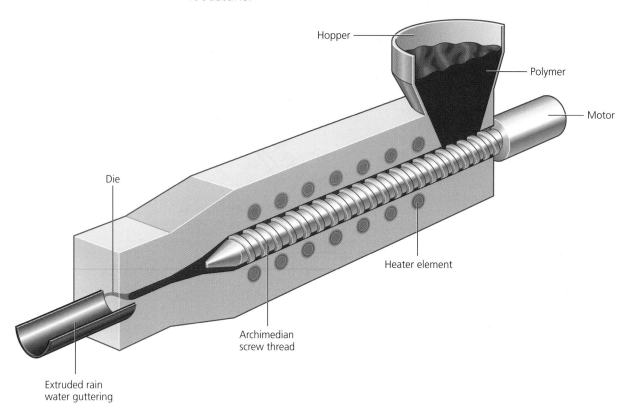

Die

Hopper

Polymer

Motor

Heater element

Archimedian screw thread

Extruded rain water guttering

Figure 3.17: PVC pipe being extruded

Plastic extrusion

Plastic extrusion starts off with plastic pellets, which are fed into a hopper where they are dried before being fed into a heating chamber. The extrusion screw forces the molten plastic through a die, forming the material into the desired shape. As the profile passes out of the die, it is often cooled in a water tank or by a mist spray.

Different plastics are used in the extrusion process to make items such as plumbing pipes, rainwater guttering, curtain tracks and uPVC window-frame sections. These are cut and welded to form whole units.

Extrusion is used to coat many products such as copper wire. For example, bare wire can be extrusion-coated with either brown or blue PVC to identify whether it is live or neutral. Extrusion coating also acts as an insulator to prevent short circuits or harm to the user. The two coated wires, together with a bare copper earth wire, can all be extruded together to form one single cable.

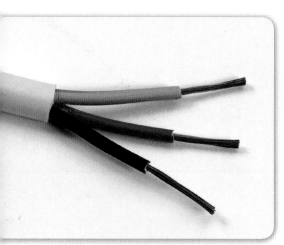

Figure 3.18: PVC-coated copper wire used for electrical installations

Figure 3.19: Plastic waste pipe extrusion

Metal extrusion

Metals can be extruded either hot or cold. Aluminium and copper are the two metals most commonly used for cold extrusion. Cold extrusion leaves no surface oxidation as a result of being heated, and a better surface finish is also achieved.

Metal extrusion is slightly different from plastic extrusion because it generally involves starting with a billet – a large lump of metal. This is heated up in an oven before being placed into the extrusion machine. A ram is then used to push the material from behind through the die.

Metal extrusion is used to form items including seamless tubes, reinforcing sections to be inserted into PVC profiles to provide greater strength to window frames, and heat sinks.

Advantages	Disadvantages
<ul><li>continuous lengths can be produced</li><li>complex profiles can be achieved</li><li>seamless tubes can be produced</li><li>small production runs can be achieved with relative ease</li><li>excellent surface finish on plastics</li><li>very high tolerances can be achieved</li></ul>	<ul><li>initial set-up costs of machinery and upkeep are high</li><li>die costs can be very high</li><li>hot extrusion of metals such as steel can leave oxidised surface finish</li></ul>

Table 3.10: Advantages and disadvantages of extrusion

Support Activity

Search the internet to find some video clips of the extrusion process. Try to find some other items such as concrete roof tiles or pasta being extruded.

Stretch Activity

Try to find a picture of a die and details of how it is made.

Joining methods

Objectives

- **Recognise** various temporary and permanent joining methods.

- **Know** how and when to use them and the preparations involved.

- **Describe** their characteristics, advantages and disadvantages.

Once the various components of a product have been made, they need to be joined together to form the complete product. The different joining techniques can be divided into two categories: temporary and permanent. You need to know about the following methods:

Temporary

- tapping and threading
- nuts, bolts and washers
- screws
- knock-down fittings

Permanent

- nails
- halving joints
- butt joints
- rebate joints
- housing joints
- mortise and tenon joints
- dowel joints
- soft soldering
- brazing
- welding
- rivets – snap and pop.

Tapping and threading

Tapping and threading can be carried out on both metal and plastics. ISO metric threads are the standard thread sizes used commercially and will satisfy most of your needs in the school workshop.

Tapping is the process of cutting an internal (female) screw thread. A hole must be drilled before the taps are used to cut the screw thread. The size of the hole must be smaller than the nominal size of the screw to allow for the cutting of the thread. An M6 thread, for example, needs a 5.0 mm diameter hole. The table below shows the tapping drill sizes.

Nominal diameter	Pitch (mm)	Tapping drill size (mm)
M4	0.7	3.3
M5	0.8	4.2
M6	1.0	5.0
M8	1.25	6.8
M10	1.5	8.5

Table 3.12: Tapping drill sizes

Support Activity

Work out what size tapping drill would be required for an M16 thread which has a pitch of 2.0 mm.

Three types of tap are used in sequence: taper, second and plug tap. The tap is held in a tap wrench, and the cutting action involves turning the tap wrench clockwise half a turn and anticlockwise a quarter turn. This action removes the swarf build-up and prevents the taps from breaking. It is essential to use a lubricant to aid the cutting process.

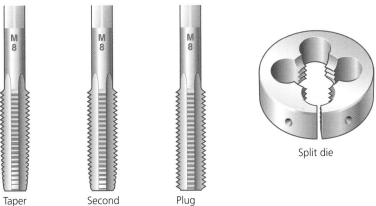

Split die

Taper Second Plug

Figure 3.20: Taps and dies

Threading is the cutting of an external (male) screw thread. A split die is held in a die stock with three screws to locate and to provide adjustments. The tapered side of the die should be used to help start the cut. The same cutting action should be used, with half a turn in the clockwise direction followed by half a turn anticlockwise to break off the swarf.

You must align the die and the rod carefully and accurately when you start threading. If the die is not square to the axis of the rod, the thread cut is classed as a 'drunken' thread.

The die is not square to the rod being cut

'Drunken' thread

Figure 3.21: A drunken thread

Results Plus
Exam Question Report

This is part of question 1 where a picture of a tool is given and candidates are asked to name and describe its use.

One of the tools in the 2009 question is shown below. (1 mark, 2009)

How students answered
Very poor set of responses, with the most common incorrect responses being file or drill. Many candidates did not answer this question.

93% 0 marks

The correct response was 'tap'.

7% 1 mark

Advantages	Disadvantages
• can be carried out in plastic and metal • size of thread can be varied slightly by adjusting the pressure on the split die • nuts and bolts widely available for use with tapped or threaded components	• both taps and dies are made from high-speed steel so they are easy to break if dropped or too much pressure is applied • taps are easily broken when threading blind holes • difficult to start threading and to make the thread parallel to the axis of the rod or hole being threaded

Table 3.13: Advantages and disadvantages of tapping and threading

Nuts, bolts and washers

Nuts and bolts provide a temporary fixing and a convenient method of securing parts that can easily be undone.

A nut is a collar, usually made from metal, with a threaded hole through the middle into which a threaded bar or bolt fits. The nut must have a thread form that matches, in size and diameter, the bolt with which it is being used. Although most nuts are hexagonal in shape, there are also square and wing nuts. A lock nut has a special nylon insert which stops it working loose.

Nut   Washer   Bolt

Figure 3.22: Nut, washer and bolt

Bolts are often made from high-tensile steel which means that they can withstand the forces applied to them. They generally have a hexagonal head at one end with a screw thread cut at the other. When using nuts and bolts, place a plain washer between the bolt head and the piece of work and another between the work and the nut. This is to protect the surface from being damaged when the nut or bolt is tightened.

As well as protecting the surface when nuts are being tightened, washers help to spread the load generated by the tightening action and to prevent the loosening action caused by mechanical vibrations. Special washers such as lock washers and spring washers also help to prevent nuts becoming loose due to vibration.

Advantages	Disadvantages
• can be undone so items can be taken apart • come in various lengths and sizes • lock nuts can be used for a firm fixing	• can work loose with vibration • if correct size is not used you can wear the head down

Table 3.14: Advantages and disadvantages of nuts, bolts and washers

Screws

Wood screws offer a reliable and neat method of fixing wood. They can be removed easily and are therefore temporary, unless they are used with an adhesive.

Screws are classified by their length, gauge, type of head and material. They are usually made from steel or brass. Brass screws are normally used where steel would rust, such as outside. Two different head types are shown in Figure 3.22 along with a cross-sectional diagram showing how two pieces of wood should be drilled in preparation for joining with screws.

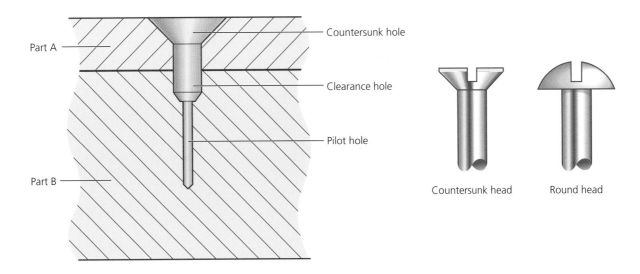

Figure 3.23: Preparation for making a screwed joint and two different screw heads

When using screws, it is important to follow these tips to avoid splitting the wood or damaging the screw head.

1 Screw through the thinner piece into the thicker one.

2 The screw should be three times as long as the piece being fixed.

3 Drill a clearance hole slightly bigger than the shank of the screw through the piece being joined.

4 Either drill a pilot hole or make one with a bradawl.

5 Use a countersink drill bit to make a countersunk recess.

6 If using brass screws in hardwood, use a steel screw of the same size first, then replace it with a brass screw.

A new range of screws has recently come on to the market, including screws which need no pilot holes, clearance holes or countersink holes. These can drill themselves in and have serrated edges on the underside of the countersink.

Stretch Activity

Work out why a steel screw must be inserted into hardwoods first if you intend to make the final joint using a brass screw.

Advantages	Disadvantages
• can be easily removed and joint taken apart if no glue has been used	• steel screws will rust if used outside
• some new types of screws do not need clearance and pilot hole drilling	• some screws can be quite hard to remove
• can be used to join dissimilar materials – plastic to wood for example	• it is difficult to get screws out if they shear off
	• if holes are not correctly prepared, screws can spilt the timber when inserted

Table 3.15: Advantages and disadvantages of screwed joints

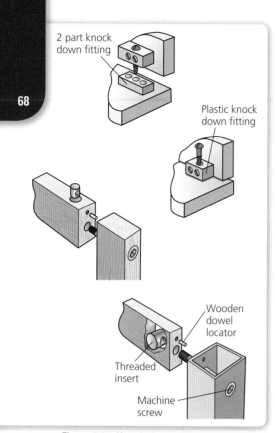

Figure 3.24: Knock-down fittings

Explain **one** advantage of using knock-down fittings when making large items of furniture. (2 marks)

■ **Basic answers (0 marks)**
A general comment about it being easier and quicker is too vague and general for any mark.

● **Good answers (1 mark)**
A statement relating to the difficulty of moving large pieces of furniture but not fully explained.

▲ **Excellent answers (2 marks)**
A fully detailed statement about the difficulty of moving large items up stairs and possibly causing damage to the piece of furniture, the decoration or the user.

Knock-down fittings

There are many modern knock-down jointing methods, all of which allow joints to be made quickly and easily. The parts can be taken apart easily and quickly, so the whole construction can be 'knocked down' or flat-packed for easy transportation or storage.

Knock-down fittings are used extensively in the construction of furniture and kitchens. Furniture is supplied flat-packed for easy transportation and assembly at home. Given that most furniture supplied in this way is made from MDF or chipboard, the fittings have to be capable of carrying and supporting weight, especially when used in kitchen wall cabinets.

Nylon joint blocks screwed into the corners, one against each face, are the most common type of knock-down fitting. These are then fastened together by a steel bolt which locks the two boards to each other. There are many variations on these such as assembly joints, a single block and a moulded single piece with a flap to conceal the screw heads.

Nails

Nails are a quick method of joining wood. As nails are driven into wood, they grip by forcing the fibres of the wood away from the head. This makes it difficult to withdraw them.

The length of the nail is important. As a general rule, it should be three times longer than the width of the wood being joined. Nails are generally sold according to their type and length.

Type	Description	Advantages	Disadvantages
round wire	made from steel wire round in section with a flat head used for general-purpose joinery 12–150 mm in length	big, flat head makes it easy to hit sometimes serrated which helps them to grip	longer nails can bend when hammering them in
oval wire	made from an oval, section wire used for fixing floorboards to joists and general joinery 12–150 mm in length	head can be punched below the surface and the hole filled	very difficult to get out as the head goes below the surface
panel pin	thinner pins in lengths up to 50 mm used for finer work such as mitre joints and small lap joints	small heads can be punched below the surface with a nail punch	they bend more easily due to their thinness

Table 3.16: Advantages and disadvantages of screw joints

Wood joints

You should always consider the choice of joint very carefully. It is important to remember that as a natural material, wood will continue to move. Joints can also be used to form features within a product, at the corners for example.

Joint type	Description	Joint
halving joints	Halving joints are made by cutting away half the thickness of the material on each half of the joint. Halving joints can be used on corners, tees, or for cross halvings. They are stronger than butt joints, and can be strengthened further by adding dowels.	
butt joints	Butt joints are the simplest form of joint and the weakest since they only have a small glueing area, which means that they can be pulled apart. They are used in cheap furniture and sometimes have dowels added to reinforce them.	
rebate joints	Rebate joints are also known as lap joints. One part of the two pieces being joined is left plain and the other has a rebate cut into it, which means that half of the thickness of the material is removed to form a lip.	
housing joints	Housing joints can be cut in natural timber and manufactured boards. They are commonly used in the construction of cabinet work for shelves or dividers.	
mortise and tenon joints	Mortise and tenon joints are widely used in the construction of furniture frames. The mortise is marked out with a mortise gauge and cut with a mortise chisel. The width of the tenon should be one-third the width of the timber.	
dowel joints	Dowel joints are butt joints with dowels used as reinforcement. Dowels are made from beech or ramin. Holes are drilled in both pieces and glue is used to secure the dowels in place and between the joining surface.	

Table 3.17: Different types of wood joint

Metal joints

Soft soldering

Soft soldering is a process for making joints in brass, copper and tinplate. When the area to be joined has been cleaned, a flux is used to prevent build-up of surface oxides and to aid the flow of solder. A thin layer of solder is applied to each of the pieces being joined. This is known as tinning. When both pieces have been tinned they can be brought together and sweated. This means a gentle flame is applied and the two tinned surfaces join to become one. Alternatively, the joint can be fluxed and heated and then small pieces of solder laid on it. This method is better when you are working with difficult and awkward shapes.

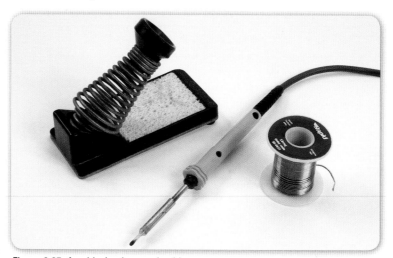

Figure 3.25: A soldering iron and solder

Brazing

When brazing, a gas-burning torch provides the heat and the flame is controlled by mixing gas and air. A flux, usually borax, is mixed with water to make a paste and is spread around the joint. The flux prevents excess oxidation and helps the brazing spelter to flow. Brazing spelter is the filler material that joins the pieces together and melts at 875°C. Brazing is therefore only suitable for use with mild steel because other metals would melt due to the temperatures involved.

Welding

Welding is a process that melts the two pieces being joined and fuses them together, so the joint is as good as the original metal. Electric arc and MIG welding are the types of welding most used in schools.

Electric arc welding uses a large electric current to jump across a small gap. With a current between 10 and 120 amps enough heat can be generated to melt the metal. A flux-coated filler rod carries the current. As it is burnt away during the welding process, the flux also burns away and protects the weld from oxidation.

MIG welding is similar to electric arc welding, but uses a continuous feed of filler rod so it does not have to be replaced. An arc is struck between the workpiece and the filler rod and an inert gas flows through the torch to prevent surface oxidation and the formation of slag.

Rivets

Rivets are most commonly used in sheet metal, although they can also be used to join acrylic and some woods to metal. They are normally made from soft iron and are available with a range of heads, the most common being the countersunk or round head. This round head type of rivet is known as a snap-head rivet.

Another form of rivet is the pop rivet, which is used with a pop rivet gun and is very useful when you only have access to one side of the object or are joining very thin sheet material. A pop rivet consists of a hollow rivet mounted on a head pin. As the head pin is drawn up through the hollow rivet by the gun, it will 'pop' when the tension on it reaches a certain point. The pop riveting process is simple and only requires a hole to be drilled for the rivet to be placed into. Although technically a permanent method of joining, both snap and pop riveted joints can be undone simply by drilling through the two pieces to remove the rivet itself.

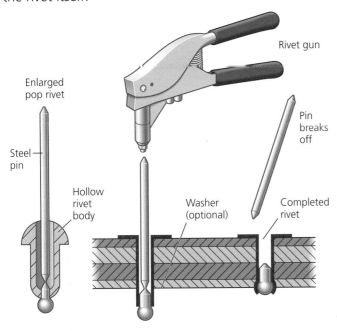

Figure 3.26: Pop riveting

Stretch Activity

Make a flowchart to show the process and stages of making a pop rivet joint.

Process	Advantages	Disadvantages
soft soldering	can be used on various metals relatively low temperatures involved	can take a long time to set up
brazing	creates a strong joint	fumes given off due to the flux burning
welding	creates strong, fused joint some devices are portable and can be easily transported	need to wear very dark goggles can only be used to join similar materials
rivets	can be drilled to undo joints joints can be created as hinges little specialist equipment required	can be time-consuming when snap riveting

Table 3.18: Advantages and disadvantages of some metal joining techniques

Adhesives

Objectives

- **Know** how to use various adhesives and the preparations involved.

- **Describe** their advantages and disadvantages.

All joints that use adhesives are permanent. It is very important that the correct adhesive is used for the specific materials being joined.

When using adhesives you should always try to make the glueing or contact area as large as possible.

You need to know about the following adhesives:

- polyvinyl acetate (PVA)
- contact adhesive
- epoxy resin
- Tensol® cement.

Polyvinyl acetate (PVA)

This is probably the most common type of woodworking adhesive. It is easy to use and apply and very strong, providing that the joints you are sticking are a good fit. The work needs to be held under pressure with clamps while the glue sets. You should remember that most types of PVA are not waterproof.

Advantages	Disadvantages
• sets in 2–3 hours	• most types are not waterproof
• excess glue easily removed with a damp cloth	
• does not stain	
• long shelf life	

Table 3.19: Advantages and disadvantages of PVA

Contact adhesive

Contact adhesives, as their name suggests, stick on contact. They are widely used for glueing large sheet materials such as thin laminates to kitchen worktops. Both surfaces must be coated with a thin layer of the adhesive, which is then left to dry in the air for approximately 15–20 minutes. It is only ready to bond when the surface is touch dry. As soon as the two coated surfaces are brought into contact, adhesion takes place and there can be no repositioning.

Contact adhesive can stick dissimilar materials together such as aluminium sheet onto MDF. Only use it in well-ventilated areas while wearing breathing apparatus as it contains dangerous chemicals.

Advantages:	Disadvantages
• sets in 20–30 minutes	• does not allow repositioning
• works well on large flat surfaces	• gives off fumes which can result in illness
• can be used to join dissimilar materials	

Table 3.20: Advantages and disadvantages of contact adhesive

ResultsPlus
Exam Question Report

Give *one* safety precaution that has to be taken when using Tensol® cement. (1 mark, 2008)

How students answered
The large majority of students were correctly able to identify one safety precaution, which was mostly related to the fumes given off or splashes to the skin.

▬	16%	0 marks
▬▬▬▬	84%	1 mark

Figure 3.27: PVA being applied to wood

Epoxy resin

Epoxy resins are versatile but very expensive. They can be used to bond almost any clean, dry materials. The adhesive is supplied in two parts. To use it, mix equal parts of resin and hardener to start the chemical reaction. Hardening of the resin starts at once, but it takes two or three days for the adhesive to reach its full strength. Epoxy resins can be used to bond dissimilar materials, but smooth surfaces need to be roughened slightly. This allows the adhesive to grip better.

Advantages	Disadvantages
• can be used to join dissimilar materials • good joint-filling properties • waterproof	• too expensive to use on large-scale work

Table 3.21: Advantages and disadvantages of epoxy resin

Tensol® cement

Tensol® cement is only used for glueing acrylic. It is a clear liquid with a solvent base that evaporates easily. It must be applied to the joint after it has been put together.

Tensol® cement works by attacking the surface it has been exposed to. It is not very strong so every effort must be made to make the glueing area as large as possible. Mask off any area you do not want the adhesive to come into contact with to avoid surface damage.

Advantages	Disadvantages
• creates a good bond between the surfaces	• gives off fumes which can result in illness • very smooth surfaces need to be roughened a bit

Table 3.22: Advantages and disadvantages of Tensol® cement

Apply it!

Many adhesives are solvent-based and give off harmful fumes. Make sure that you only use them in well-ventilated areas and avoid contact with your skin. Always read and take note of the manufacturer's instructions and warnings.

Support Activity

Look at a tin of Tensol® cement or contact adhesive and draw the symbols pictured on the labels. Find out what each of the different symbols means.

Heat treatment

Objectives

- **Know** how to use various heat treatments to alter the properties of metals and the preparations involved.

- **Describe** their characteristics, advantages and disadvantages.

The term heat treatment describes the process of heating and cooling metals in a controlled manner. This allows changes to be made to the properties of the metal, such as increasing its hardness or reducing its brittleness.

You need to know about the following heat treatments:

- hardening and tempering
- annealing
- case hardening.

Hardening and tempering

Increasing the hardness of steel is only possible where the steel contains more than 0.4 per cent carbon. The full effects of hardening are only possible where the carbon content is over 0.8 per cent. It is necessary to increase the hardness of steel used to make tools such as scribers, drills and punches. Tools like these must therefore be made from medium-carbon steel or silver steel.

To harden a piece of silver steel fully, heat it to just over 720°C which can be judged as a dull red colour. Soak it at this temperature until it is uniformly heated and then quench it immediately in water or oil.

At this point, the steel is very hard but too brittle to be of any practical use. The hardness needs to be reduced slightly to produce a more elastic, tougher material that will retain a cutting edge. This is done through tempering.

To temper a piece of hardened steel, first clean it with emery cloth or wire wool. This cleaning allows the colour of the oxide formed on the surface to be seen easily and clearly as the component is reheated.

Heat the steel at a point well behind the cutting edge. As the metal heats up, an oxide appears on the surface and you will see this oxide passing along the metal as it gets hotter.

When the appropriate colour reaches the cutting edge or tip, it should be quenched immediately in water. The hardness and brittleness are both reduced as the temperature increases.

Approx temperature (C)	Colour	Toughness	Uses
230	pale straw	least	lathe tools
240	straw		scribers
250	dark straw		centre punch
260	brown		tin snips
270	brown-purple		scissors
280	purple		saw blades
290	dark purple		screw driver
300	blue	most	springs

Table 3.23: Tempering colour chart

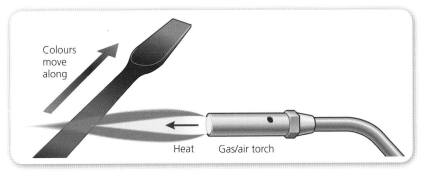

Figure 3.28: The tempering of a screwdriver blade

Annealing

As a metal is worked or deformed by bending, rolling, or hammering, its structure changes. Its hardness increases as a result of this cold working, and it becomes more difficult to work. To ease this, the material needs to be annealed. The annealing process restores the initial structure of the material by relieving internal stresses.

The process of annealing involves heating the metal to a certain temperature depending on the material, and then allowing it to cool. It is quite common practice to dip or 'pickle' cooled brass and copper in dilute sulphuric acid. This process chemically removes the scale that forms on the surface during the annealing process. Once annealed, the material is soft again and can be further worked and shaped.

Case hardening

The only way to harden mild steel that does not contain much carbon is to case harden it. This gives it a carbon-rich skin while keeping the more elastic, ductile properties on the inside.

Heat the metal to a cherry red, dip it into a carbon powder, and leave it there to cool. This process should be repeated three or four times; each time increases the metal's hardness since it is during the reheating process that the carbon is absorbed into the material's surface. The final step is to heat it again to a cherry red and quench it in water.

This process is useful for producing tool holders and driveshafts that require a hard outer surface.

Process	Advantages	Disadvantages
hardening and tempering	allows the hardness of a material to be increased for specific purposes	once hardened most products need to be tempered product has to be cleaned so that the colours can be seen cannot harden steel which has a carbon content below 0.4%
annealing	allows materials to be softened which will enable further deformation to take place	surface produces a scale when being heated and cooled which is difficult to clean
case hardening	only the external surface can be hardened, leaving the centre core still quite soft in comparison	process has to be repeated several times in order to make a significant difference

Table 3.24: Advantages and disadvantages of heat-treatment processes

Finishing techniques

There are many finishing processes and techniques for woods, metals, and plastics. Plastics do not generally require a great deal of treatment and surface finishing because of the nature of the material and the manufacturing techniques involved.

You need to know about the following finishes:

- varnish
- wax polish
- stain
- paint
- plastic dip-coating
- electroplating.

Finishing and surface treatments are usually carried out to improve the following:

- aesthetics – the appearance of the material
- functional properties – protect the material from deteriorating and to prolong its useful life.

The suitability of a finishing process depends on:

- the material used
- where it is going to be used.

Consider the choice of finish carefully in the early stages of a design. You must also consider its durability and the aftercare required. Maintenance is essential if a product or component is expected to last for many years. Generally, finishes are applied before final assembly. This is especially important on internal surfaces.

Before any finish can be applied, wood and metal surfaces must be prepared. Plastics do not often undergo major surface treatments.

Preparing metal surfaces

Ferrous metals need to be prepared thoroughly regardless of the finish. Remove all surface oxides with emery cloth or wet and dry paper. If you are going to paint the surfaces, you need to degrease them first using methylated spirits on a rag. Figure 3.29 shows two pieces of metal, one that has come from the store cupboard and one that has been carefully cleaned ready for painting.

Figure 3.29: One piece of metal has been correctly prepared and the other has not

Preparing wood surfaces

Wood must also be prepared carefully before any finish can be applied. You can use a plane to produce a clean, smooth surface. Remove any minor surface blemishes with progressively finer grades of glass paper. Prevent any scratching by working along the grain. You can use an electric sander on larger flat surfaces.

Figure 3.30: Electric sanders for use in the school workshop

Varnish

Synthetic resins (plastic varnishes) produce a much harder, tougher surface than some of the natural finishes such as shellac. They are heatproof and waterproof and fairly resistant to knocks.

You can buy varnishes in a wide range of shades and finishes such as matt or gloss. Varnish is best applied in thin coats with a brush or spray. Gently rub down each layer with wire wool after you have applied it, although this does make the process very time-consuming. You also need to be very careful with the dust in the workshop as it can stick to wet surfaces.

Always apply varnish in the direction of the grain, in light, even strokes as shown in Figure 3.31.

Figure 3.31: Varnish being applied to a table top

Wax polish

Wax oil produces a dull gloss shine. It is made from beeswax which has been dissolved in turpentine to form a paste and is applied to timber using a cloth. The addition of silicon wax or carnauba wax greatly increases the durability of the wood.

Before applying any wax, you need to seal the surface of the timber. You can use shellac, a natural resinous product dissolved in either cellulose or methylated spirits, which penetrates and seals the surface.

You might use a wax finish on a wooden desk, coffee table or dining room table where you want to show off the natural grain and colour of the timber, although this type of finish is prone to marking if someone places a hot cup on the surface.

Stain

Staining or colouring enhances the natural grain of timber. It is very much a decorative finish and allows for an even application of colour.

Stains can be water-based, spirit-based or oil-based. Spirit-based stains tend to dry more quickly, whereas oil-based stains last longer and are more versatile. Both spirit- and oil-based types are flammable and they must both be used in a well-ventilated area due to their toxic nature.

Stains are available in a wide range of colours and can also be used to mimic other woods. You can use colours such as oak and mahogany on cheaper softwoods to give the impression of a more expensive product.

You can brush stains on or apply them with a cloth. In industry, products are either sprayed or dipped, which allows a much quicker and more even application.

Paint

Both woods and metals can be painted. Paint is used to provide a decorative colouring and protective layer whether used indoors or outside.

Painting wood

When preparing wood to be painted, seal any knots with knotting to prevent any resin from seeping through and spoiling the appearance. Gently round off any sharp corners with glasspaper. Then seal the wood with a primer. After rubbing it down gently, apply an undercoat. Finally, apply a top coat. Most paints for wood are either oil- or polyurethane-based. Both are durable and waterproof but the polyurethane type is generally much tougher.

Quick notes

Finishing processes are used for three main reasons: to enhance the aesthetics of the material, to protect the material or surface, and to prolong the lifespan of the product.

Painting metal

Preparing metals before painting is very important. Surface oxides must be removed and surfaces degreased. A red-oxide paint is generally used as a first coat to prevent further oxidation of the surface.

Apply a primer and undercoat before the final top coat. Rub the surface down gently between each layer with some very fine wet and dry paper.

In industry, metals are normally sprayed in water chambers to take away the smell and the waste material. In the motor industry, car spraying is now widely carried out by robots, as can be seen in Figure 3.32. They have been programmed to follow the shape and contours of the car very carefully. Spraying cars this way results in a much better quality of finish.

Hammerite is a particular type of metal paint that does not need extensive surface preparation. It can be painted straight onto metal almost regardless of the surface condition. It is typically used on wrought-iron gates and fences, and old-fashioned cast-iron drainpipes. It is also often used on workshop machinery and it can easily be identified by its 'cracked' textured surface appearance. It can be quite difficult to apply, although there is a spray can alternative which is a little easier to use.

Plastic dip-coating

Dip-coating is suitable for most metals. It is used for coating products such as hanging baskets, brackets, kitchen drainers and tool handles.

The metal must first be thoroughly cleaned and degreased before being heated in an oven to 180°C. Soak it at this temperature before plunging it quickly into a bath of a fluidised powder. Leave it there for a few seconds while the powder sticks to the hot surface to form a thin coating. Then return the object to the oven, allowing the plastic coat to fuse to leave a smooth, glossy finish. The major disadvantage of this type of finish is that it can crack and peel off over time.

Electroplating

Electroplating is often used to give metals such as brass and copper a coating of a more decorative durable metal such as silver or chromium. This means that a cheaper base metal can be used with a more expensive material coating over the top. The process is carried out by electrolysis: the product is charged and the solution acts as a conductor. The process is quite expensive and it can take quite a long time to create a layer of the required thickness.

Figure 3.32: Robot paint sprayer

Manufacturing processes for batch production

Objectives

- **Know** how and when to use jigs and patterns when manufacturing products and components, and the preparations involved.

- **Describe** their characteristics, advantages and disadvantages.

In school, you will design and make one-off products on your own. In industry, where many thousands of the same product may have to be made, more efficient ways of making things have to be found.

Sometimes this principle will apply to your work: for example, if you are batch-producing a number of identical components such as shelves. To do this you need to develop a more efficient use of time as well as a method to ensure that the same level of quality is maintained.

Jigs and patterns

In this kind of work, you need accurate and fast marking out. Jigs and patterns are one way of achieving this in industry and in school workshops.

Jigs

A jig is a work-holding device that is made specifically to suit a single component or a device to help when making identical components. The component is held firmly in the exact position required. The jig is not clamped to the work table, but is free to be positioned and held against the workpiece being cut, either by hand or with clamps.

The jig shown in Figure 3.33 is a very basic one that could be made in a school workshop. You would use a jig such as this to make dowel joints to join boards or sheets together to make a piece of furniture. The jig would be marked out and drilled once and then held against the boards to drill through, saving time and repeated marking out.

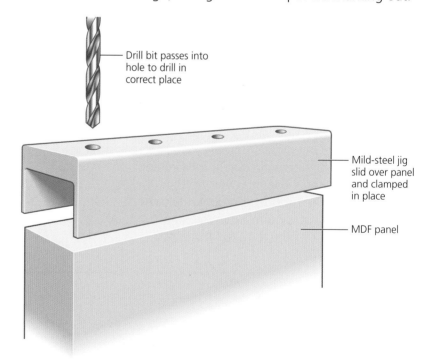

Drill bit passes into hole to drill in correct place

Mild-steel jig slid over panel and clamped in place

MDF panel

Figure 3.33: A drilling jig for use when making dowel joints

The jig pictured in Figure 3.34 can be used to cut dovetail and finger joints. Dovetails take a long time to mark out. They are very difficult to cut, and take a great deal of skill and practice. Jigs like the one pictured can be used with a router to cut very accurate joints over and over again in the same place. When making several drawers or a small wooden box, for example, the router jig is an ideal solution.

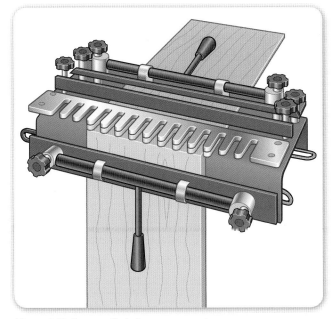

Figure 3.34: A router jig being set up, ready to use

Support Activity

Search the internet to watch someone cutting dovetails using a router jig.

Patterns

A pattern is used to mark out or make a number of identical components. A pattern or template is often made from metal and is used by marking or drawing around its edges. Patterns are particularly useful when marking out complex and difficult shapes.

Sand casting patterns are used when casting, and have sand packed around them. They are an exact copy of the finished product and are often made from wood or an epoxy resin with a very good surface finish. Patterns can be used repeatedly for batch production. They can also be made as a one-off, for example to replace a broken component.

Process	Advantages	Disadvantages
jig	very useful when making many identical components or products much quicker than marking out lots of identical items	must be set up and aligned correctly, or all the holes or joints will be cut in the wrong places can take some time to set up
pattern	can be reused to make a batch of products	can take a long time to make the pattern highly skilled workers are required to make patterns

Table 3.25: Advantages and disadvantages of using a jig or a pattern

Health and safety

Objectives

- **Understand** and be able to describe safe working practices.

- **Identify** workshop hazards and precautions.

Support Activity

Look at the photograph below and make a list of safety points and rules which are being applied by pupils and staff.

In 1974, the Health & Safety Executive (HSE) was formed to ensure that companies and organisations conform to specified health and safety standards. Its aim is to safeguard the health, safety and welfare of employees and to safeguard others, especially the general public.

Health and safety regulations cover a wide range of areas including temperature, lighting, noise, special clothing, disposal of waste materials and chemicals, and machine and plant maintenance.

The school workshop and environment are governed by strict health and safety legislation. You must follow safe working practices in a safe working environment when you are at school.

Safe working practices

- Keep your work area clean, tidy and well organised.
- Keep the area between benches and machines clear to avoid potential hazards which could cause you to trip and fall.
- Check your tools before use to make sure that they are in good, safe condition. Report to your teacher any that are blunt or broken.
- Report any faults, breakages or damage so that things can be repaired or replaced.
- Keep both hands behind the cutting edge when using sharp tools such as chisels.

Figure 3.35: A typical school workshop

Most accidents in workshops are caused by carelessness. The best way to create a safe working environment is to follow some basic guidelines on health and safety. Make sure you know:

- what to wear
- how to behave
- how to keep your working environment safe
- how to work safely
- the accident procedure.

What to wear

Loose clothing of any sort is potentially dangerous. Remove all jewellery before using any machinery. Take off your jacket or blazer, tuck in your tie and put on an apron or a workshop coat. If you have long hair, tie it back. Wear good, strong footwear. You should also wear special eye protection when necessary. Always put safety items and equipment back for others to use when you have finished with them.

How to behave

Accidents in a school workshop are often the result of silly behaviour. Behave sensibly at all times and do not distract others, especially when they are working on machines. Do not rush around the workshop. This is particularly important if you are carrying tools and materials.

A safe working environment

It is very important to be aware of potential hazards in the workshop – electrical, heat, chemical or dust. Use personal protective equipment (PPE) when you are exposed to dust, chemicals or any form of heat treatment. PPE includes gloves, breathing apparatus or masks, ear defenders and eye protection.

Always check electrical equipment, such as soldering irons, drills and sanders, before use to make sure that it is safe and that there is no frayed, burnt or exposed flex.

When you have been working in the heat treatment area or casting bay, leave hot tools and work in a cool place before putting them away. You should also let others know that this area is hot.

Many modern adhesives and paints have a chemical base. Read the warnings on the can and take notice of the advice. Glass-reinforced plastic (GRP), for example, must only be used in a well-ventilated area and you should wear a mask while you are using it. Tensol® cement and contact adhesives should also be used in a well-ventilated area. Always wash your hands after you have finished working and tidying away.

Safety symbols in the school workshop

- Mandatory instructions are shown in blue.
- Hazard warnings are shown within a diamond shape.

Figure 3.36: Workshop warning signs

Know Zone
Chapter 3 Industrial and commercial processes

Industrial and commercial processes focus specifically on manufacturing techniques and processes associated components. Forming, joining, heat treatment and finishing processes are central to any manufacturing, be it for one-off, batch or mass production.

You should know...

☐ the characteristics, applications and the advantages and disadvantages of:
  ☐ one-off production
  ☐ batch production
  ☐ mass production
☐ processes relating to a range of materials and processing and forming techniques and their respective advantages and disadvantages when joining materials and components
☐ temporary and permanent joining techniques and their respective advantages and disadvantages when joining materials and components
☐ adhesives, including their advantages and disadvantages when joining materials
☐ heat-treatment processes including their advantages and disadvantages and how they can be used to alter the properties of metals
☐ finishing techniques and how they are used to improve the performance, quality and appearance of manufactured products
☐ how to use jigs and patterns when manufacturing products and components
☐ how to describe safe working practices
☐ how to identify workshop hazards and precautions.

Key terms

hardening and tempering   annealing

                              case hardening

Which of the key terms best fits each description below?

A restores the initial structure of the material by relieving internal stresses

B increases brittleness

C makes the product harder

D makes the outside surface harder than the core inside

To check your answers, look at the glossary on page 173.

Multiple-choice questions

1. What type of component is shown in the diagram below?

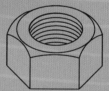

A nut   **C** washer

B bolt   **D** screw

2. Which of the following finishes can be applied to a metal?

A varnish   **C** wax

B plastic dip-coating   **D** stain

ResultsPlus
Maximise your marks

The handle of the screwdriver is plastic dip-coated.

Explain *one* reason for plastic dip-coating the handle of the screwdriver. (2 marks, 2007)

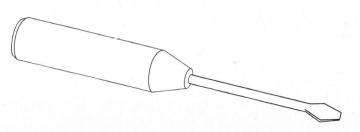

Student answer	Examiner comments	Build a better answer
● It will make it easier to grip (1 mark)	This is a good start for 1 mark but the response fails to explain fully what benefit this will have.	△ The grip will be improved which makes it easier to hold when undoing tight screws.
● The plastic will not let electric through it meaning you will not get hot (1 mark)	Another correct statement although the correct terminology – 'insulator' – has not been used. The term has also been misapplied making reference to the handle getting hot rather than acting as an insulator from electric shocks.	△ The plastic coating will act as an insulator which means you will not get an electric shock.
■ Makes it look nice (0 marks)	The reference here is to aesthetics, which is correct, but it makes no statement about how it achieves this.	△ The plastic comes in a variety of colours which means that a colour can be applied, making it more aesthetically pleasing.

Overall comment: In very many cases, students failed to recognise any benefits of plastic dip-coating the handle. Very few gained full marks for a detailed explanation.

Practice exam questions

1. (a) Name the type of joint shown in the diagram below to join the rail to the table leg. (2)

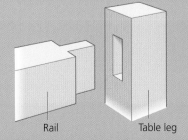

Rail   Table leg

(b) The leg is made from oak.

Name **one** surface finish that can be applied to the table leg. (1)

(c) Describe **one** advantage of applying a surface finish to the table leg. (2)

Chapter 4 Analysing products
Specification criteria

Objectives

- **Analyse** a product taking various specification criteria into account.

When you analyse a product, it is a good idea to look at it carefully and then judge or assess it against the criteria listed below.

- **Form** – why is the product shaped/styled as it is?
- **Function** – what is the product for?
- **User requirements** – what makes the product attractive to potential users?
- **Performance requirements** – what does the product need to do and how can this be achieved technically?
- **Material and component requirements** – what materials and components are needed and how should they perform?
- **Scale of production and cost** – how does the design allow for scale of production? What do you need to consider to determine cost?
- **Sustainability** – how does the design take environmental considerations into account?

Analysing products helps you understand how different products function. This helps you to recognise ideas that work well and disregard other ideas and products that are less effective.

Case study: a hand-soap dispenser

Soap dispensers are used in many bathrooms and kitchens.

Figure 4.1: A hand-soap dispenser

The product shown in Figure 4.1 has been analysed against the following criteria.

Form

- Water will run off the bottle due to its tapered shape.
- It is smooth, which makes it easy to hold.
- It has a large base which makes it quite stable.
- The lever at the top is large which makes it easy to push down.

Function

- Each time the lever is depressed, a set amount of liquid soap is dispensed.

User requirements

- It is made from plastic which is easy to clean.
- As a set amount of soap is dispensed, very little is wasted.
- It is see-through, so it is easy to see how much soap is left inside.

Performance requirements

- When the lever has been pushed down, it should return to the up position.
- It should be capable of being locked closed so that no soap can come out during transportation.

Material and component requirements

- The spring mechanism should be reliable and work every time.
- The material should be easy to clean as it may become dirty when people use it.

Scale of production and cost

- The bottle container is a simple shape which can be blow-moulded.
- The other plastic pieces are complex shapes but they can be manufactured by injection moulding in large volumes.
- Although the initial cost of the moulds is high, unit costs will be relatively low.

Sustainability

- The plastic pieces are all thermoplastics, which means that they can be recycled.

There are many versions of this type of product. Before designing a new and improved version, the product designer or company will probably carry out a product analysis similar to the one above, looking at which features work well and which do not in a wide range of similar products. Some soap dispensers are designed specifically to appeal to young children to encourage them to wash their hands more frequently. These dispensers often have a novelty feature like the one in Figure 4.2, which is in the form of a fish.

Support Activity

Carry out a product analysis on a pair of scissors for cutting paper. You might find it easier to lay your work out in a table.

Figure 4.2: A novelty hand-soap dispenser designed and made to appeal to a younger audience

Materials and components

Objectives

- **Identify** the materials and/ or components used in manufacturing a product.

- **Understand** their properties, qualities, advantages and disadvantages.

- **Justify** the choice of materials and/or components.

Choice of materials is critical when designing and manufacturing new products or components. You must be aware of the specific properties and qualities of the material you have chosen.

Case study: a garden spade

A garden spade, like the one shown below, is made from several materials, each with its own properties.

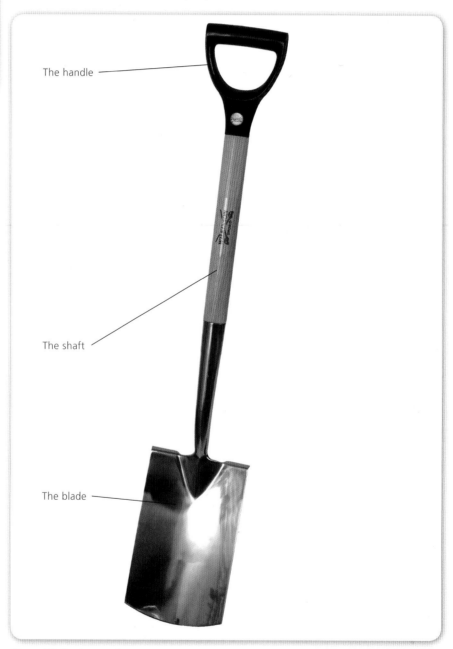

The handle

The shaft

The blade

Figure 4.3: Garden spade

The garden spade consists of three separate parts: the blade, the shaft and the handle. Each part is made from a different material.

The blade

The blade is made from stainless steel.

Properties and qualities

- Stainless steel is very hard and tough, which is essential as the blade will be used to cut into hard soil and possibly through roots. Due to its hardness, the blade can be ground with a sharp edge.

- It has excellent resistance to corrosion, which is also very useful as it will be used in wet soil and is likely to be stored in damp conditions, such as a garden shed or garage.

- The excellent surface finish means it can be easily cleaned. Its high lustre means that it needs no surface finishing other than a quick polish and buff.

- However, stainless steel is a difficult material to use and join, and specialist welding equipment is required.

The shaft

The wooden shaft is made from ash, a hardwood.

Properties and qualities

- Ash is tough, so it will be able to absorb some of the impact as it is thrust into the soil.

- Its elastic properties will enable it to withstand some bending as it is levered back during digging.

- The shaft is finished with a clear varnish, which makes the wood more waterproof and durable.

The handle

The handle is injection-moulded from ABS.

Properties and qualities

- ABS is a tough plastic with high-impact strength and this makes it an ideal choice for this component.

- It is also lightweight and durable, which means that it is easier to carry and long-lasting.

- As it is injection-moulded, very little surface finishing is required other than perhaps removing any flashing or sprue pins.

- It can be coloured by adding a pigment at the moulding stage.

- However, ABS is quite expensive compared with some other thermoplastics, although the expense can be fully justified given its properties and many advantages.

ResultsPlus
Exam Tip

You will notice in the text how correct terminology and property definitions have been used rather than generic terms such as 'strong'. It is important to learn about the properties and applications of materials so that you are able select and justify the use of a material in a particular situation.

Support Activity

Use the internet to find out how a garden spade is manufactured.

Manufacturing processes

Objectives

- **Identify** the processes involved in the manufacture of products, including the stages of the manufacturing process.

- **Understand** their advantages and disadvantages.

- **Justify** the choice of manufacturing process.

When you are designing a product or component it is important to have a clear idea of how you are going to manufacture it. The same product can usually be manufactured in a variety of ways. Specialised machinery and tooling will sometimes be required.

Case study: a screwdriver

Screwdrivers are one of the most common tools, in homes and garages as well as in workshops. The screwdriver shown in Figure 4.4 is presented in a blister pack ready to be sold in shops and DIY stores. The stages of production involved in getting the product to this point are outlined below.

Figure 4.4: A blister-packed screwdriver

1. The steel bar is forged to shape the blade end.

2. The blade end is heat-treated, which improves its properties by hardening and tempering it.

3. The whole bar is electroplated.

4. The handle is injection-moulded and formed around the steel bar to hold it firmly.

5. The two-part blister pack is vacuum-formed.

Each manufacturing process is further described and justified in Table 4.1.

Stage	Process	Justification of each process
1	Forging	This is a quick process once the tip has been heated to the correct temperature. A forming tool has been cut to the correct shape so that the screwdriver tip and blade are identical every time.
2	Heat treatment	Many blades are heat-treated at the same time, making the whole process cost-effective. The temperature is carefully controlled to achieve an exact degree of hardness. The blades have to be cleaned before they can be tempered, which can be time-consuming and costly. The cost of energy needed to reach the high temperatures required makes this process quite expensive.
3	Electroplating	This process involves dipping the blades into the plating solution through which an electric current is passed; it works like electrolysis. It can take a long time to build up a relatively thick surface layer, so it can be quite expensive and time-consuming. However, the result is an excellent, durable surface finish.
4	Injection moulding	This process is expensive to set up since the machinery and tooling cost a lot at the outset. However, the unit costs are low, and identical components can be manufactured 24/7. The main advantage is that the blades can be moulded directly onto the handles. The same handle and mould can also be used to fit different-sized screwdriver blades, which makes the whole process even more economically viable.
5	Vacuum forming	The transparent packaging that contains the screwdriver is vacuum-formed over a simple mould. This process also involves a single mould, which can be used repeatedly 24/7. The process can be automated and is relatively quick, given the size of the package.

Table 4.1: Reasons for choice of manufacturing processes to make a screwdriver

Support Activity

Using the diagram of an exploded biro in Figure 4.5, identify the processes used to manufacture the cap and ink tube. Justify your choice of manufacturing processes along with the advantages and disadvantages of the two selected processes.

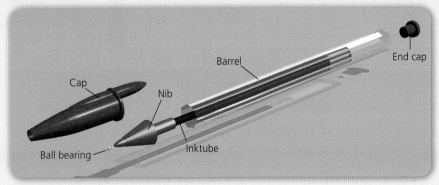

Figure 4.5: Exploded view of a biro

exam zone

Know Zone
Chapter 4 Analysing products

When you are trying to design a new or improved product, product analysis is a good place to start. It is also useful when trying to develop an understanding of why certain materials and processes have been used in a particular product.

You should know...

- [ ] how to analyse a product taking into account a specific set of criteria which includes:
  - [ ] form
  - [ ] function
  - [ ] user requirements
  - [ ] performance requirements
  - [ ] material and component requirements
  - [ ] scale of production and cost
  - [ ] sustainability
- [ ] how to identify materials and or components used in the manufacture of a product including:
  - [ ] the properties and qualities of the materials and or components used
  - [ ] the advantages/disadvantages of the materials and components used
  - [ ] how to justify the choice of the materials and or components used
- [ ] how to identify the processes involved in the manufacture of products including:
  - [ ] how to justify the choice of manufacturing processes used
  - [ ] the advantages/disadvantages of the manufacturing processes used
- [ ] how the product meets the specification points given
- [ ] how to compare and contrast two similar products using the same criteria.

Key terms

form

function

user requirements

performance requirements

material and component requirements

scale of production and cost

sustainability

Which of the key terms best fits the description below?

Why is the product shaped or styled as it is?

To check your answers, look at the glossary on page 173.

Practice exam questions

1. A bicycle frame shown in the diagram below is made from mild steel.
   (a) Give **two** properties of mild steel that makes it suitable for the bicycle frame. (2 marks)

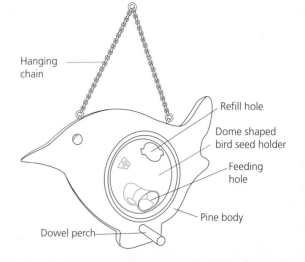

★ ResultsPlus
Maximise your marks

The dome-shaped birdseed holder is made from acrylic.

Give *one* property of acrylic that make it suitable for the birdfeed holder.

(1 mark, 2009)

Hanging chain

Refill hole

Dome shaped bird seed holder

Feeding hole

Pine body

Dowel perch

Student answer	Examiner comments	Build a better answer
■ **Strong** (0 marks)	This term is too general since strength is better described or classified as tensile, compressive or shear strength.	△ Good impact strength is a good response since it relates to the fact that it will withstand small pecks from birds.
● **Waterproof** (1 mark)	A good relevant property.	△ A good property since the product will be outside and it will need to withstand the rain.
● **Plasticity** (1 mark)	Another good relevant property.	△ A good property and one which relates to the material's ability to be heated and shaped in order to make the dome shape required.

Overall comment: A large majority of candidates demonstrated a general lack of knowledge relating to the properties of acrylic.

Chapter 5 Designing products
Specification criteria

94

Designers create new products in response to a need, a demand and new technologies, or simply to improve and develop an existing product.

Figure 5.1: The Alessi lemon squeezer has become a design icon. It is an example of a design created because it is aesthetically pleasing, rather than to improve function.

Designers often start by looking at, observing and analysing existing products. You learned about the criteria used to judge products in the previous topic. The same criteria are used when designing new products.

- **Form** – why is the product shaped/styled as it is?
- **Function** – what is the product's purpose?
- **User requirements** – what makes the product attractive to potential users?
- **Performance requirements** – what does the product need to do and how can this be achieved technically?
- **Material and component requirements** – what materials and components are needed and how should they perform within the product?
- **Scale of production and cost** – how does the design allow for scale of production? What do you need to consider to determine cost?
- **Sustainability** – how does the design take environmental considerations into account?

In your written examination paper you will be asked to design a product that meets a given set of criteria. You will need to produce two separate design ideas that meet the stated requirements. The specification points given will be taken from the specification criteria listed above.

ResultsPlus
Watch out!

To get full marks for the design question, you must produce two different design ideas that address all eight specification points differently. For example, you cannot use the same materials and processes in each idea.

Design exam question

You have been asked to design a trophy for a table tennis competition.

The specification for the table tennis trophy is that it must:

- show that it is for the winner
- show that it is for a table tennis competition
- show the name of the winner
- allow for the winner's name to be changed each year
- provide a stable base
- not damage the surface on which it is placed
- be made from materials available in a school workshop
- be manufactured using processes available in a school workshop.

Design ideas

Below, you can see two different solutions from a student in response to the question and specification points given above.

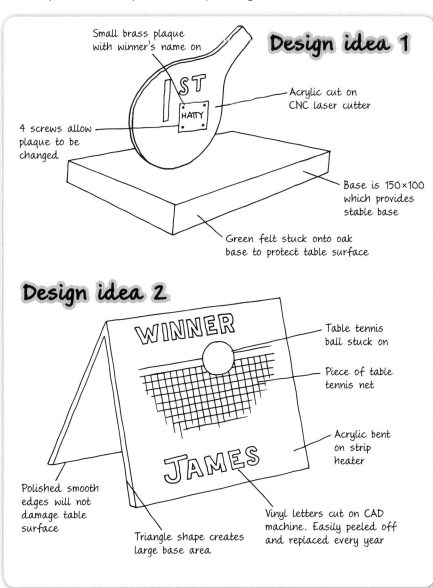

Design idea 1

Small brass plaque with winner's name on

Acrylic cut on CNC laser cutter

4 screws allow plaque to be changed

Base is 150×100 which provides stable base

Green felt stuck onto oak base to protect table surface

Design idea 2

WINNER

JAMES

Table tennis ball stuck on

Piece of table tennis net

Acrylic bent on strip heater

Polished smooth edges will not damage table surface

Triangle shape creates large base area

Vinyl letters cut on CAD machine. Easily peeled off and replaced every year

Figure 5.2: A student's design ideas

Results Plus
Watch out!

Specification points can be covered in the form of brief notes, rather than by trying to show everything graphically. So a statement saying 'made from acrylic and cut on the laser cutter' will score 2 marks, 1 each for the material and the process.

Support Activity

Try answering the design question above in no more than 20 minutes.

Designing skills

When you answer the design question, you must communicate your ideas clearly. Simple outline drawings with clear notes are the best way to do this. If you use pencil for the diagrams and sketches, it must be a dark pencil such as HB or B.

When annotating your design ideas, try not to write too much, but do try to be specific and use the correct terminology. Be specific about materials and processes and avoid general terms like 'wood', 'metal' and 'plastic'.

One simple drawing supported by detailed annotation is often enough to address all the specification points in the question.

Figure 5.3, which shows the work of a student in response to a set design question, is a good example of this. The original question is also shown below.

Design question

Design a table-top system to display tourist information leaflets.

The specification for the table-top system is that it must:

- hold two different-sized leaflets
- not fall over
- hold at least 50 leaflets
- stop the leaflets falling out
- display the leaflets clearly
- allow easy access to the leaflets
- be made from materials available in a school workshop
- be manufactured using processes available in a school workshop.

Additional information

Leaflet 1 measures 300 mm high by 100 mm wide.

Leaflet 2 measures 150 mm high by 150 mm wide.

A stack of 50 leaflets has a thickness of 25 mm.

Design idea 1

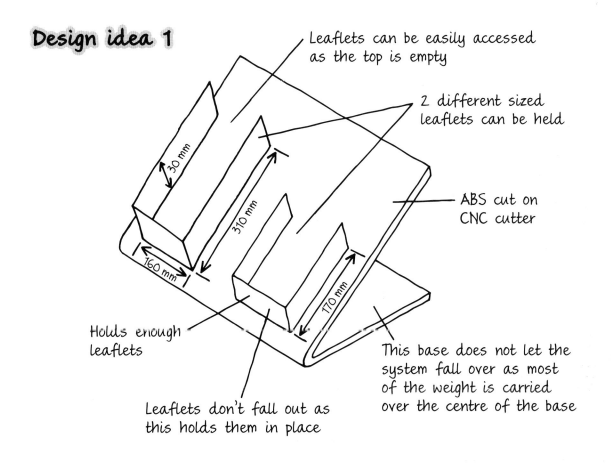

Leaflets can be easily accessed as the top is empty

2 different sized leaflets can be held

ABS cut on CNC cutter

30 mm

310 mm

160 mm

170 mm

Holds enough leaflets

Leaflets don't fall out as this holds them in place

This base does not let the system fall over as most of the weight is carried over the centre of the base

Design idea 2

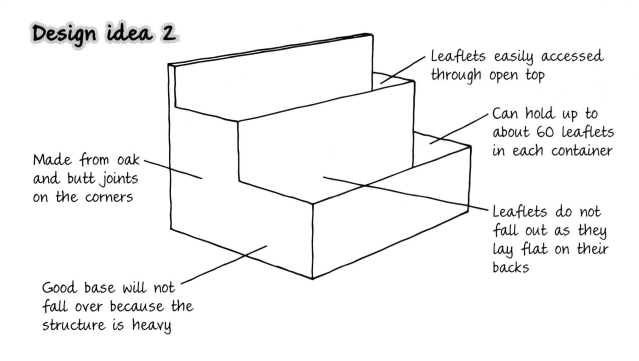

Leaflets easily accessed through open top

Can hold up to about 60 leaflets in each container

Made from oak and butt joints on the corners

Leaflets do not fall out as they lay flat on their backs

Good base will not fall over because the structure is heavy

Figure 5.3: A student's design ideas

Application of knowledge and understanding

Objectives

- **Apply knowledge and understanding** of the properties of a wide range of materials and/or components and manufacturing processes to design ideas.

- **Apply knowledge and understanding** of the advantages and disadvantages of materials and/or components and manufacturing processes to design ideas.

- **Justify** the choice of materials and/or components and manufacturing processes.

When designing a product, first sketch out some rough ideas and concepts. As you start to improve and develop your ideas, you will need to apply your knowledge and understanding of materials, their properties and how they can be used to manufacture products.

Be aware of the properties of materials when you are considering using them for specific products or components. For example, if you need a component hard enough to withstand indentation and abrasive wear, copper would not be suitable because it is too soft, whereas something like carbon steel would be much more appropriate.

The diagram in Figure 5.4 shows a bird feeder designed by a student. The sketch is fully annotated to give details of materials, properties and an outline of the manufacturing process that would be used for its construction.

Support Activity

Make a table showing two properties for each of the materials listed below:
- oak, mahogany and pine
- mild steel, aluminium and copper
- acrylic, PVC and ABS.

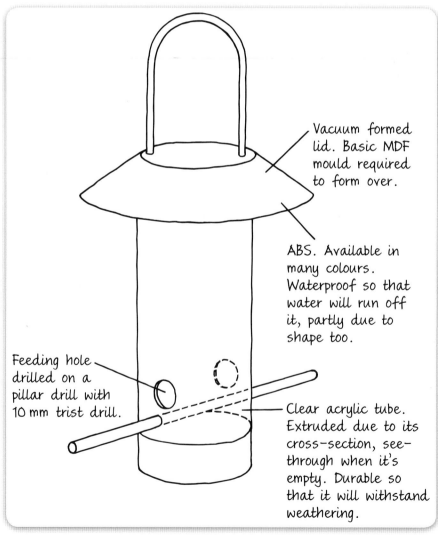

Vacuum formed lid. Basic MDF mould required to form over.

ABS. Available in many colours. Waterproof so that water will run off it, partly due to shape too.

Feeding hole drilled on a pillar drill with 10 mm trist drill.

Clear acrylic tube. Extruded due to its cross-section, see-through when it's empty. Durable so that it will withstand weathering.

Figure 5.4: A student's design for a bird feeder

The photograph in Figure 5.5 shows a pair of workshop pincers used to remove nails from wood.

Figure 5.5: A pair of workshop pincers

If you are choosing materials for this product, you need to be aware of the forces involved at the jaws and on the riveted joint, and you should think about the type of surface finish that could be applied to the handles.

Carbon steel has been used because it is hard. The jaws should be able to withstand any damage from the steel nails on which they will be used. They also need a ground edge so they can grip the nails they will be pulling out.

A rivet has been used to join the two separate pincer pieces, allowing the pieces to move. The rivet must be capable of withstanding the forces to which it will be subjected when the pincer pieces are used to grip nails.

The handles have been plastic dip-coated. This will give the user a better grip when applying a force to the two handle parts, as well as covering any imperfections to provide a smoother surface finish.

Stretch Activity

Apply your knowledge and understanding of the materials involved in making the soldering iron shown in Figure 3.25. The tip of the soldering iron is made from copper. The electrical cable has been PVC coated. List their advantages and disadvantages for this particular product and justify their use. You might like to present your work in the form of a table.

Know Zone
Chapter 5 Designing products

Designing products allows you to demonstrate your creative skills and ability. It also gives you an opportunity to apply your knowledge and understanding of materials, properties and manufacturing processes.

You should know...

☐ how to respond creatively to a design brief and set of specification criteria

☐ how to clearly communicate your ideas

☐ how to use notes and sketches

☐ how to use annotation that relates back to the specification criteria

☐ how to apply your knowledge and understanding of materials, components and manufacturing processes to your design ideas.

Key terms

form

function

user requirements

performance requirements

material and component requirements

scale of production and cost

sustainability

Which of the key terms best fits the description below?

What qualities would make the product attractive to potential users?

To check your answers, look at the glossary on page 173.

Practice exam questions

1. Design a bathroom cabinet.

The specification for the bathroom cabinet is that it must:

☐ be fixed to a wall

☐ provide a mirrored surface

☐ be waterproof

☐ store a bar of soap

☐ hold four toothbrushes

☐ allow the toothbrushes to be easily removed

☐ be made from materials available in a school workshop

☐ be manufactured using processes available in a school workshop.

Use sketches and, where appropriate, brief notes to show two different design ideas for the bathroom cabinet that meet the specification points above. (16)

ResultsPlus
Maximise your marks

You have been asked to design a table-top system to display tourist information leaflets.

The specification for the table-top system is that it must:

- ☐ hold two different sized leaflets
- ☐ not fall over
- ☐ hold at least 50 leaflets
- ☐ stop the leaflets falling out
- ☐ display the leaflets
- ☐ allow easy access to the leaflets
- ☐ be made from materials available in a school workshop
- ☐ be manufactured using processes available in a school workshop.

Additional information

Leaflet 1 measures 300 mm high by 100 mm wide
Leaflet 2 measures 150 mm high by 150 mm wide
A stack of 50 leaflets has a thickness of 25 mm
(16 marks, 2003)

Student answer	Examiner comments	Build a better answer
Back is dipped to hold leaflets more efficiently / *Easy to take leaflets* / *Steps to show different kinds of leaflets* / *Plywood* / *Sturdy base* (3 marks)	The candidate has not provided sufficient details. There is a lack of specific detail about manufacturing processes. No use has been made of the dimensions to show where the leaflets would go or to indicate how they would be stored or removed.	Greater use of dimensions would have helped, along with a little more annotation to detail what processes could have been used. Details about how the leaflets were to be removed would also have helped, rather than simply stating 'easily removed'.
ABS plastic– see through / *Lots of space to pull leaflet out* / *Two different sized leaflets* / *25 mm* / *Wide stand, metal steel, strong, robust* (7 marks)	A very good response on the whole. Well drawn, clear and concise annotation. Use of dimensions would have helped a little but the candidate has drawn the two different-sized leaflets in their respective holders.	The only element this candidate has missed is that a manufacturing process has not been named. This could have been easily added with some annotation.

Overall comment: On the whole, there was a good range of answers from many candidates but the scores tended to be better for the first design idea. The scores were lower for the second design idea because candidates did not differentiate it enough from the first design idea. In many cases, the materials and processes were the same for both ideas, and so no marks could be awarded for this element for the second idea.

Chapter 6 Technology Information and communication technology (ICT)

Objectives

- **Understand** the role of ICT in the design, development, marketing and sale of products.

- **Describe** its advantages and disadvantages.

- **Know** how it affects society.

Information and communication technology (ICT) is used to communicate and transfer data from one system or user to another. Communication systems have become very advanced and continue to improve all the time. Networks now exist in almost every environment where there are computers, such as coffee shops, railway stations and airports. Manufacturers use ICT to keep track of products, components, accounts, sales data and production control.

Email

Email means that designers, manufacturers, retailers and consumers can communicate much more quickly – and cheaply – than they used to. Emails are transmitted at very high speeds and can reach a computer on the other side of the world in a matter of seconds. Pictures and other types of computer file including databases, text files, video clips and spreadsheets can also be sent by attaching them to an email message; the recipient clicks on the attachment to open it. This means that companies can send files quickly to their clients. Designs, prices and production schedules can all be created in various software packages and saved as separate files, which can then be emailed as attachments to clients anywhere in the world. Clients can respond immediately in the same way, saving time and money.

Use of email in marketing and sales

Online retailers use email to market and sell their products. They do this by sending emails about new products, special offers and promotions, and so on, to customers who have registered on or purchased from their site. This gives them direct access to customers who like their products or services. Sometimes companies will sell personal details to other marketing companies that they think may be interested in their products, leading to further email traffic.

Advantages	Disadvantages
• very high speed	• lots of spam (junk email) generated which slows down the internet and clogs up servers
• large address books can be compiled for mass-mailing	
• orders and payments can be made	• hackers can get into accounts and gain access to your computer, files, and personal information
• eliminates postage costs	
• mobile technology means that email can be received and sent on the move	• you do not know whether the person you are communicating with is actually who they say they are

Figure 6.1: Shoppers have constant access to email accounts via wi-fi

Table 6.1: Advantages and disadvantages of email

Electronic point of sale (EPOS) systems

Electronic point-of-sale (EPOS) systems are used to gather and record information, for example in shops, when items are scanned at the checkouts. Each product in a shop will have its own unique barcode that is either printed on it or on a label attached to the product. This barcode is read by a laser scanner.

The barcode consists of a series of black stripes on a white background. The stripes and the spaces between them vary in width. A number is printed below the barcode in case the strips get damaged and cannot be read by the laser scanner. Retail codes usually use a 13-digit number that gives information about the product's country of origin and manufacturer, as well as its specific product code.

The EPOS system has been described as the 'intelligent' till in supermarkets. As the product is passed over the laser scanner, the barcoded item is identified by its unique code and its sale is recorded on a computer. The computer monitors sales across the whole store. It also records the current stock levels of each item, and reorders stock from the distribution centre as needed.

Companies can use the information gathered by EPOS systems to analyse sales and stock levels since all the data is stored on a computer. The system also records and works out how much money the shop has made, giving the company more control over theft, wastage and damage.

Use of EPOS type systems in manufacturing

Manufacturing industries are adopting stock control principles similar to those used by the EPOS system in their own production and stock control systems. During the manufacturing process, products are given a unique barcode just like a product in a supermarket. Indeed, some companies have developed their own coding systems for specific components that are used in the assembly of a product.

Fixings such as screws or pop-rivets are boxed into a batch of, say, 50 and then barcoded. This process is repeated for all the individual components needed to manufacture the product. When boxes are booked out for assembly, they can be tracked throughout the whole assembly process along with all the other components. Replacements are ordered automatically since stock levels are recorded.

This recording and tracking process allows companies to monitor product assembly much more carefully with respect to quality control and quality assurance. Another advantage is that they do not have to keep large stocks of bought-in components, which means that they can control spending on stock more easily.

The EPOS system has many benefits for supermarkets and manufacturing industries, including:

- quick and efficient sales and order processing
- barcode search makes checking stock levels quick and easy
- ability to adjust and record stock levels on a daily basis
- can generate daily reports on, for example, sales history
- easy to keep customers' and suppliers' details, including what your favourite items are at the supermarket.

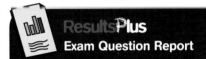

| Origin of product | Brand owner's number | Item number | Check digit |

Figure 6.2: Barcodes are used on every product

Describe *two* advantages for the manufacturer of using an EPOS machine that reads the barcode. (4 marks, 2008)

How students answered

Some students did not answer this question, which shows a very limited level of knowledge or in fact no knowledge at all.

29%   0 marks

A good number of students were able to make a comment about being able to track products or to record stock levels, but were unable to go on to fully describe the advantage.

66% 1–3 marks

Only very few students were correctly able to give an advantage, such as stock being able to be quickly located in a warehouse.

 5% 3–4 marks

Digital media and new technology

Objectives

- **Understand** the transfer of data using Bluetooth wireless personal area networks, its uses, advantages and disadvantages.

- **Understand** the uses of videoconferencing, its advantages and disadvantages.

Digital media and new technologies have expanded electronic communication, which is now possible on the move. Technology allows meetings to take place between designers and clients who might be on different sides of the world.

Bluetooth® wireless personal area networks

When you use your computer, games console or telephone the various pieces and system parts connect with each other via wires and cables. As systems have become more complex, technology such as Bluetooth® has been developed, which allows some of these connections to be made via radio waves.

Bluetooth® is an open wireless system that uses radio waves to exchange data over short distances between mobile and fixed devices. This data exchange takes place over a personal area network (PAN).

Bluetooth® allows you to connect and exchange data between laptops, PCs, mobile phones, digital cameras and games consoles such as Wii and PlayStation 3. Devices used in Bluetooth® PANs must be fitted with the appropriate transmitters and receivers, such as the mobile phone headset shown in Figure 6.5.

Figure 6.4: The Bluetooth® symbol

Figure 6.5: A typical mobile phone Bluetooth® headset worn by a driver

Stretch Activity

If you have any Bluetooth® devices, test them to see what their range is.

Advantages	Disadvantages
• low power consumption so can run for a long time on a small battery	• games controllers rely on battery power and can run out mid-game
• devices do not need to be in line of sight to work because they use radio waves	• services can be interrupted by power failure which can result in data being lost in transfer
• higher powered devices can work up to a range of 100 metres, so systems can be installed in places such as railway stations and shopping malls, providing customers with network coverage	• security can also be a concern for Bluetooth® devices. There are two main security problems for Bluetooth®: Bluejacking and Bluebugging.

Table 6.2: Advantages and disadvantages of Bluetooth® wireless PAN

Videoconferencing

A videoconference, sometimes also called a video teleconference, is a communication system that allows two or more people in different locations to have an interactive video and audio conversation.

Telecommunications equipment is used to relay pictures and sound around the country or to other countries, allowing virtual meetings to take place and avoiding the need for costly, time-consuming journeys.

Figure 6.6: A typical videoconference set-up where groups of people can all see and hear each other

A videoconference requires some basic equipment such as:

● video input;   video camera or webcam
● video output;   computer monitor, television or projector
● audio input;   microphone
● audio output;   loudspeakers
● data transfer;   LAN, digital telephone line or internet.

This technology allows designers to hold discussions with their clients without having to travel.

Advantages	Disadvantages
• saves time (no travelling needed) • saves money (no travel expenses) • better for the environment, as no need to travel long distances • more efficient because many people from different locations can be involved • allows employees to work from home which cuts down travelling	• can be quite expensive to set up all the necessary technological equipment and resources • internet speed can vary and may be lost, making it difficult to retain any sense of flow • some users do not feel comfortable talking to a camera and microphone

Table 6.3: Advantages and disadvantages of videoconferencing

Support Activity

Describe **two** advantages of using videoconferencing as a teaching method if schools are closed due to adverse weather conditions.

Computer-aided design/computer-aided manufacturing (CAD/CAM) technology

Objectives

- **Understand** what CAD/CAM is.

- **Understand** the uses, advantages and disadvantages of virtual modelling and testing; laser cutting; computer numerically controlled (CNC) milling and turning; rapid prototyping.

Developments in computer technology have benefited designers and manufacturers enormously. Designers can use computer-aided design (CAD) systems to create, develop, record and communicate with others anywhere in the world. Computer-aided manufacturing (CAM) systems translate design data into codes, and programs into manufacturing data, enabling computer numerically controlled (CNC) machinery to cut or turn products and components automatically, quickly and accurately.

Virtual modelling and testing

A virtual model is a computer (or digital) model of a physical object. Virtual modelling is used mainly for visualisation purposes.

Virtual models can also be tested using simulation programs. Air flow, stress analysis and fatigue testing can all be carried out in a virtual laboratory. The image in Figure 6.7 shows how testing can be used. It is much safer and cheaper to simulate this type of crash than to use real cars and people.

Figure 6.7: Virtual testing of a crash against a crash barrier

Advantages	Disadvantages
• products can be coloured, and textures added to show what they will look like in real life • designs can be changed easily without redrawing the whole image • files can be sent electronically via email to clients and manufacturers, which saves time and money • electronic files can be linked to CAM machines so that prototypes can be manufactured	• software can be expensive to purchase • learning to use the software can take a long time

Table 6.4: Advantages and disadvantages of virtual modelling and testing

Laser cutting

Laser cutting is a developing technology that uses a high-powered laser controlled by a computer to cut various materials. The laser works by either melting or burning away the material. It is best used on a flat surface, although it can be used to cut round materials. The finished edge requires little surface finishing. This type of technology is becoming increasingly available in schools. Figure 6.8 shows a student using it to cut out parts for a project.

Figure 6.8: Laser cutter being used to cut sheet metal

Advantages	Disadvantages
• highly accurate and capable of achieving fine detail	• initial capital outlay on the machine is high
• can work 24 hours a day, 7 days a week since it is fully automated	• lasers can damage the eyes
• easy to cut identical components	• not very effective on highly polished mirrored surfaces

Table 6.5: Advantages and disadvantages of laser cutting

Support Activity

Use the internet to find a video clip of a laser cutter in action.

Computer numerically controlled (CNC) milling and turning

CNC machinery can be used to cut and turn products and components automatically with great accuracy and speed. A milling machine uses a cutter that moves up and down vertically while the work which is fixed to the table moves backwards, forwards and from side to side. When carefully controlled, the milling machine can cut curves and 3D shapes and profiles.

CNC lathes are used to turn products in the round. As the work is held in the chuck, the tool post moves to turn cylindrical products. In some instances, tool changes are automatic and all the normal turning processes can be carried out. Some CNC lathes have pneumatic chucks and automatic material-feeding devices, so non-stop production is possible.

Figure 6.9: A CNC lathe

Advantages	Disadvantages
• can work 24 hours a day, 7 days a week	• initial capital outlay on the machine is high
• extremely accurate; identical copies are produced each time	• CNC milling and turning requires highly specialised staff
• fewer manual workers required as labour costs are lower	
• complex shapes and forms can be achieved	

Table 6.6: Advantages and disadvantages of CNC milling and turning

Rapid prototyping

Rapid prototyping is a process of automatically creating physical objects by adding materials in layers to build up a 3D object. It takes a virtual design from a CAD package and converts it into thin horizontal layers that sit on top of each other until the model is complete.

Rapid prototyping is commonly used in the automotive and aerospace industries to produce prototypes quickly for testing and demonstration purposes.

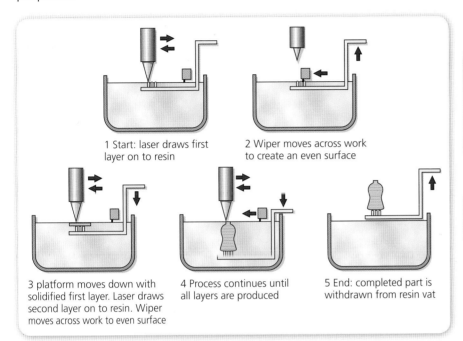

1 Start: laser draws first layer on to resin

2 Wiper moves across work to create an even surface

3 platform moves down with solidified first layer. Laser draws second layer on to resin. Wiper moves across work to even surface

4 Process continues until all layers are produced

5 End: completed part is withdrawn from resin vat

Figure 6.10: One method of rapid prototyping being used to produce a 3D model

Advantages	Disadvantages
• prototypes can be produced very quickly from a design idea	• initial capital outlay on the machine is high
• full 3D complex-shaped products or components can be formed	• models are sometimes very fragile and break easily
• a 3D model is easier to handle than a 2D image, and communicates a design idea more quickly	
• products can be developed, tested and brought to market more quickly	
• fewer manual workers are required so labour costs are lower	

Table 6.7: Advantages and disadvantages of rapid prototyping

New developments in computer technology mean that designers can use powerful software to create exciting designs that can be modelled and tested before manufacturing. ICT developments allow faster and more efficient communication between designers, clients, manufacturers, retailers and consumers.

You should know...

- [ ] about the role of ICT in the design, development, marketing and sale of products
- [ ] about the effects of ICT on society
- [ ] about the advantages and disadvantages of ICT
- [ ] how email is used between designers, manufacturers, retailers and consumers
- [ ] about EPOS
- [ ] how the internet can be used for marketing and sales
- [ ] about the use of digital media
- [ ] about Bluetooth® wireless personal area networks
- [ ] about videoconferencing
- [ ] about the advantages and disadvantages of digital media
- [ ] what CAD/CAM is
- [ ] about virtual modelling and testing
- [ ] about laser cutting
- [ ] about computer numerically controlled (CNC) milling and turning
- [ ] about rapid prototyping.

Key terms

EPOS   laser cutting

Bluetooth®   CNC

videoconference   rapid prototyping

virtual modelling

Which of the key terms best fits the description below?

A wireless personal area network.

To check your answers, look at the glossary on page 173.

Multiple choice questions

1. Which of the following is a disadvantage of using email?

A Very high speed of data transfer.

B Cuts down on postage costs.

C Lots of spam mail generated.

D Large address books can be compiled.

2. Which of the following is an advantage of using a Bluetooth® wireless headset?

A Service can be interrupted by power failure.

B They use small batteries, which run out.

C They need to be plugged into a computer.

D They do not need to be used in line of sight of the transmitter or receiver.

3. Rapid prototyping is a process that:

A automatically creates a physical 3D object.

B is very labour intensive.

C takes a long time to produce a prototype from a design idea.

D can only produce 2D, flat models.

4. Which of the following is a disadvantage of using a laser cutter?

A It is a very accurate process.

B It can achieve very fine detail.

C It can cause damage to your eyes.

D It can operate 24 hours a day, 7 days a week.

ResultsPlus
Maximise your marks

A new design for a pair of wooden steps has been sent electronically by the designer to the manufacturer.

Explain *one* advantage for the designer of being able to send the new design electronically to the manufacturer.
(2 marks, 2008)

Student answer	Examiner comments	Build a better answer
■ It is electronic. (0 marks)	This response is not even worth 1 mark because it basically restates information that has been given in the question. There is also no further explanation.	△ The files will be in an electronic format which means that they can be loaded straight into CAM machines. (2 marks)
● Faster than posting. (1 mark)	This is an advantage but since it is not fully explained to say what the benefits are, it can only be awarded 1 mark.	△ Faster than post because it only takes a few seconds to send a message when using email. (2 marks)
● Can be seen anywhere in the world faster. (1 mark)	An advantage providing that there are electronic links and services where it needs to be sent but no further detail is given.	△ Can be sent around the world quickly because messages are sent down telephone lines and via satellites rather than by overland postal services. (2 marks)

Overall comment: On the whole quite well answered for 1 mark but candidates did not always go on to explain the advantage they had chosen fully.

Practice exam questions

1. Bluetooth® data transfer relies on which of the following? (1)

   A radio waves

   B infrared

   C copper cables

   D microwaves

2. (a) Name **two** pieces of basic equipment necessary to hold a videoconference. (2)

   (b) Describe **one** advantage of being able to have a videoconference between a designer and a client. (2)

Chapter 7 Sustainability
Minimising waste production

Objectives

- **Understand** the four Rs of minimising waste production: reduce materials and energy; reuse materials and products; recover energy from waste; recycle materials and products and use recycled materials.

- **Describe** their uses, advantages and disadvantages.

Support Activity

List five products that could be reused in different ways.

Waste minimisation is about trying to reduce the amount of waste produced by individuals and society in general. Manufacturing processes should be designed to keep resource consumption and energy use to a minimum. Product designers must consider the product life cycle, what happens to the product and how is it disposed of when it reaches the end of its useful working life.

Reduce materials and energy

More efficient manufacturing processes and better materials are being developed. Minimising waste often requires additional investment due to new, efficient machinery, but savings can be made as a result because they use less energy. More efficient use of materials can also lead to minimising waste. Lighter and stronger materials have led to a reduction in the size of some structural components, which means that less material is required overall in order to make a given product: for example, a bike fork made from carbon fibre.

Using fewer components in a product reduces the amount of material used. It also makes the product easier to take apart and recycle at the end of its useful life.

Manufacturers are reducing energy consumption in manufacturing, for example by turning off lights at night, in an effort to strike a balance between a product's life expectancy and the amount of energy needed to manufacture it.

Reuse materials and products

Reusing means using a material or product more than once. A product might be used repeatedly in the same way, like a milk bottle, which is collected from the doorstep, washed, and refilled with milk, or in a different way, such as an old bathtub used in the garden as a planter.

In the case of the milk bottle, reusing the product means that new material is not required to make replacements. Reusing the bathtub means it has not been taken to the landfill site.

Figure 7.1: Milk bottles are cleaned and used again

Recover energy from waste

Energy can be recovered from waste (EFW – energy from waste) and waste can be converted into energy (WTE – waste to energy), in the form of either heat or electricity. In some locations, this heat is fed directly into domestic houses to meet their heating needs, which means that the demand on natural resources such as gas and oil is reduced.

Both EFW and WTE involve incinerating waste, which means burning it to dispose of it but also to produce useful energy. The amount of energy recovered depends on the efficiency of the incineration plant. Release of the emissions generated by the incineration process must be carefully controlled and monitored and such emissions must be cleaned up or 'scrubbed' before they can be discharged into the atmosphere.

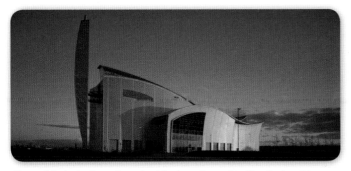

Figure 7.2: An incineration plant that produces heating for local houses

Recycle materials and products and use recycled materials

Recycling means processing used material and products into new materials or products. The aim is to reduce the demand for new materials, reduce energy consumption and reduce air pollution. This means that less waste is produced as fewer products or items are being disposed of at waste or landfill sites.

Glass, paper, metal, textiles and some plastics can all be recycled. Although these materials are not typically recycled or reprocessed into new, pure products, they are reused in the manufacture of different materials such as cardboard or lower-grade metals and plastics.

Salvage is another type of recycling. It involves stripping out materials such as gold from old mobile phones or computers, and lead from car batteries. This is only generally carried out where the salvaged materials have a high value, as with gold.

Figure 7.3: Recycling centre with bays for different materials

Exam Question Report

Explain *three* benefits to the environment of recycling household waste. (6 marks, 2008)

How students answered

This part was poorly done with very few students being able to give benefits.

20% 0–1 marks

This part was poorly done with very few students being able to give anything more than a basic response, such as less landfill space is required.

56% 2–3 marks

Some students were able to offer reasons such as less waste going to landfill, less mining for new material and a reduction in energy consumption, and were also able to go on to explain the benefits for the environment.

24% 4–6 marks

Support Activity

Try to find out how much waste is disposed of every year in the UK and how much is recycled. Try to find out what costs are associated with the savings that are made by recycling materials.

Energy sources

Objectives

- **Understand** the uses, advantages and disadvantages of wind energy.

- **Understand** the uses, advantages and disadvantages of solar energy.

- **Understand** the uses, advantages and disadvantages of converting biomass into biofuels for use in vehicles.

Renewable energy is energy generated from natural resources that cannot run out, such as the sun and the wind. Biofuel, a liquid fuel, is another form of renewable energy derived from biological materials, or biomass, such as dead trees, branches or wood chippings.

Wind energy

When the wind blows, the moving airflow can be used to turn wind turbines. In recent years, the efficiency of turbines has increased, and most modern wind turbines can generate up to 5 megawatts of power. Wind turbines are positioned where there is a good wind flow, such as offshore and on the tops of hills. A wind farm is a collection of wind turbines that have been grouped together like those shown in Figure 7.4.

Solar energy

Solar energy refers to the whole concept of collecting sunlight and using the sun's energy. Solar energy can be used to generate electricity, to heat water and buildings, and even to cook in solar ovens.

A solar cell converts the sun's energy into electricity using the photovoltaic effect, which involves the creation of a voltage, due to the cell being directly exposed to the sun's radiation.

Figure 7.4: An offshore wind farm

Support Activity

Use the internet to find out where onshore and offshore wind farms are located.

Stretch Activity

Try to find out how much of the global energy demand is met by wind, solar and biomass sources of energy.

Figure 7.5: A photovoltaic power plant in Spain

Using more solar energy will reduce the need to generate electricity by other means, such as coal-burning power stations, reducing harmful emissions. Photovoltaic power stations like the one shown in Figure 7.5 are capable of producing millions of watts of power. When run in association with other forms of renewable energy, they will contribute to meeting an increasing global demand.

Biomass

Biomass is generally used in power plants as a combustible fuel. Wood is a biomass material and has been used for thousands of years as a heat source and for cooking. Biomass also includes biodegradable wastes that can be burnt as fuel.

Field crops are also used for biomass, and some crops are grown specifically for conversion into biofuel. With some crops that are grown for food, the by-products can be used as a fuel source. Biofuels are made by fermenting the sugar component parts of the plant to produce ethanol. Although ethanol can be used as a fuel on its own, it is often mixed with conventional fuels to increase the octane and to reduce vehicle emissions.

Renewable energy source	Advantages	Disadvantages
wind energy	does not create greenhouse gas during operation long-term potential suggests wind power could produce five times total current global energy demand	high initial installation costs unsightly noisy large amount of land required
solar energy	can be placed out of sight on roof tops once installed and paid for all energy is free requires little maintenance can be installed along roadsides acting as a noise barrier	expensive to install supply can be reduced by cloud cover less useful in countries with fewer sunshine hours
biomass	leftover parts of food crop can be used	large amount of land required to produce sufficient crops not ideally suited for fuel burning wood produces soot

Table 7.1: The advantages and disadvantages of renewable energy sources

Stretch Activity

Find out why biomass is not ideally suited for transportation fuel.

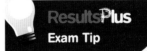

ResultsPlus
Exam Tip

Make sure that you learn about the advantages and disadvantages of all three forms of renewable energy.

Climate change

Objectives

- **Understand** how developed countries are responsible for minimising the impact of industrialisation on global warming and climate change.

- **Understand** how the Kyoto Protocol aims to reduce greenhouse gas emissions.

Figure 7.7: Polar ice-caps are melting as a result of global warming

Support Activity

Find out how much each of the following contribute to the greenhouse effect: carbon dioxide, methane, water vapour and fluorinated gases.

Global warming is the gradual increase in the average temperature of the earth's atmosphere due to the greenhouse effect (see below). Climate change caused by global warming can be seen in changing average weather patterns and in an increase in extreme weather events, such as the floods that caused devastation in many areas of the UK in the summer of 2007.

Figure 7.6: In 2007, flooding caused havoc in many areas across the UK

Effects of global warming

Global warming is causing the polar ice-caps to melt, and sea levels to rise as a result. Rainfall patterns are changing noticeably. Glaciers are melting at record rates, and extreme weather events around the globe are becoming more common. Significant loss of agricultural yields and further loss of animals, with some species becoming extinct, is expected as a result of changing conditions within ecosystems.

Gases present in the atmosphere, such as carbon dioxide, methane, water vapour and fluorinated gases, act like a greenhouse around the earth. This means that they allow the heat from the sun into the atmosphere, but do not allow it to escape back into space. These gases are called greenhouse gases.

Scientists believe that rising levels of greenhouse gases in the earth's atmosphere are creating an increase in global temperatures that will have potentially harmful consequences for the environment and human health. Human activities, such as burning fossil fuels and deforestation, are thought to be causing the levels of these gases to rise.

Kyoto Protocol

The Kyoto Protocol is an international agreement linked to the United Nations Framework Convention on Climate Change, aimed at reducing greenhouse gas emissions globally. It was originally adopted in Kyoto, Japan on 11 December 1997 and came into effect on 16 February 2005. As of Spring 2010, there were 84 signatories and 190 parties to the Kyoto Protocol.

The Protocol sets targets for 37 industrialised countries and the European Community (EC) to reduce emissions of the four greenhouse gases, carbon dioxide, methane, nitrous oxide and sulphur hexafluoride. These countries have agreed to reduce their greenhouse emissions to at least five per cent below 1990 levels. However, emissions from air travel and international shipping are not included.

In order to ensure that the Kyoto Protocol is being implemented, five key principles were established.

1 The commitment to reduce greenhouse gases was legally binding.
2 Countries had to prepare policies and measures to reduce greenhouse gases.
3 A fund for climate change would be set up to help minimise the impact on developing countries.
4 Reports and reviews of the Protocol would be carried out.
5 A committee would be established to ensure that the Protocol is being complied with.

At the time of inception, countries like China, India and other developing countries were not included because they were not considered to be the main contributors to greenhouse gas emissions. Despite this, and given that China has since become one of the biggest greenhouse gas producers, some developing countries also made a commitment to share this international common responsibility to reduce emissions.

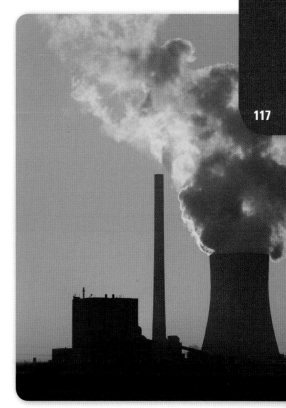

Figure 7.9: Emissions need to be controlled in an attempt to cut greenhouse gases

Support Activity

Make a list of the possible environmental consequences of not cutting greenhouse emissions.

As manufacturing industries have increased output in order to meet consumer demand, there has been an ever-increasing demand on the world's natural resources, many of which are non-renewable. There is an increasing emphasis on minimising waste production and sustainable design.

You should know...

☐ about the principles of minimising waste production throughout the product life cycle
☐ how to apply the 4Rs
☐ about the use of renewable energy sources
☐ about the responsibility of developed countries for minimising the impact of industrialisation on global warming and climate change.

Key terms

reduce

reuse

recover

recycle

photovoltaic cells

biomass

Kyoto Protocol

Which of the key terms best fits the description below?

To use a material or product more than once.

To check your answers, look at the glossary on page 173.

Multiple choice questions

1. Which of the following is NOT true about waste minimisation?

A Reducing the amount of waste produced by society

B Reducing the amount of people working in a factory

C Reducing the amount of energy consumed in the factory

D Reducing the amount of materials used in manufacturing

2. Which of the following is a form of renewable energy?

A Oil

B Gas

C Coal

D Solar

3. Which of the following is an advantage of using solar energy?

A It is expensive to install.

B It is very labour intensive.

C It takes a long time to produce a prototype from a design idea.

D It can only produce 2D, flat models.

ResultsPlus
Maximise your marks

Each year thousands of items are not being recycled.

Explain *one* reason why items are *not* being recycled.
(2 marks, 2009)

Student answer	Examiner comments	Build a better answer
● They just get thrown away (1 mark)	This response is just about worth one mark and no more since there is no explanation.	▲ Some materials cannot be recycled because it costs too much to do so therefore they just get thrown away into landfill sites. (2 marks)
● Sorting waste takes too much time (1 mark)	A good reason which is just about explained.	▲ Having to sort the waste into different types such as paper, plastic and glass tales too much time. (2 marks)

Overall comment: On the whole quite well done for 1 mark but candidates did not always go on to fully explain the advantage stated.

Practice exam questions

1. Which of the following is **NOT** a method of minimising waste?

    A Reduce

    B Recycle

    C Recover

    D Remove

    (1)

2. Renewable energy sources are being used more to generate electricity.

    (a) Describe **two** disadvantages of using wind farms in a small village. (4)

    (b) Describe **two** advantages of using solar cells to generate electricity. (4)

Chapter 8 Ethical design and manufacture
Moral, social and cultural issues

Objectives

- **Understand** the uses, advantages and disadvantages of built-in obsolescence when designing and manufacturing new products.

- **Understand** the uses, advantages and disadvantages of offshore manufacturing of mass-produced products in developing countries.

- **Understand** the importance of respecting different cultures when designing and manufacturing products.

Figure 8.1: Electrical products being made and assembled in China

Support Activity

Make a list of five products that have a limited lifespan and explain why you think this is the case.

As well as the practical problems involved in product design and manufacture, you need to consider various ethical, social and moral issues.

Built-in obsolescence

There is a tendency towards a 'throwaway' culture – when products stop working or break, you just throw them away. In design and engineering terms, some products are designed to stop working after a certain period of time. This is known as built-in obsolescence.

Built-in obsolescence is in evidence in many different products such as computer software and hardware, where companies will stop providing support and spare parts for older versions in an effort to make consumers buy new updated versions. Manufacturers sometimes use inferior or less expensive components in their products to ensure that they only work for a certain period of time.

Advantages of built-in obsolescence

- When a product stops working, the consumer often buys a new one. This means the manufacturer continues to make money.
- New, improved and more expensive products are gradually brought to the market, so the consumer often replaces an outdated product with a newer, more expensive version. This means the manufacturer makes more money.

Disadvantages of built-in obsolescence

- Sometimes consumers buy replacement products from a different manufacturer, fearing that one from the same manufacturer may break again.
- Consumers have to spend more money on replacing products or upgrading software.
- Broken or out-of-date products often end up in landfill sites.

Offshore manufacturing

Offshore manufacturing is the practice of large manufacturing companies and industries relocating their businesses from one country to another. It has become widespread in recent years because companies have been able to take advantage of cheap labour costs available in some countries.

China has become one of the biggest offshore manufacturing countries following its accession to the World Trade Organisation in 2001. As global communications have improved, India's manufacturing capability has also increased.

Advantages of offshore manufacturing

- Companies can use cheaper labour rather than pay the higher wages in the UK, or another developed nation.
- Fewer and cheaper costs, for example energy and labour costs, than in the UK due to local practices and prices.

Disadvantages of offshore manufacturing

- Loss of jobs and workforce in home countries of businesses.
- Transporting products around the world results in high energy costs and creates pollution.
- Loss of secondary jobs, which are based on providing parts and suppliers to the relocated business.

Respecting cultures

As designers create new products, they must be aware of current and future trends and market needs. They must also consider cultural, social, moral and environmental issues in relation to the product and where in the world it is going to be used.

Designers are very concerned with the aesthetics of their products. However, what is aesthetically pleasing to one person is not necessarily aesthetically pleasing to another. Designers must also consider whether the product is appropriate for where it is going to be used and for the people who will be using it.

People of different cultures have different views and opinions on what is aesthetically pleasing. Often these views are based on historical and religious images and works. It is important for designers to be aware of both the cultural and the artistic influences on their work as well as the overall design for manufacture. For example, some graphic images used in advertising may cause offence to certain groups of people and cultures. Many consumers, often the young ones, are concerned with the image that a product gives them. With this in mind, designers face increasing pressure to create products that convey a particular image.

Stretch Activity

Produce a photo board or collage of how mobile phone technology and products have been designed over the past 25 years. Make some notes about their size, form and functions.

Figure 8.2: The music industry is a leader in developing cultures and trends associated with certain fashions

exam zone

Know Zone
Chapter 8 Ethical design and manufacture

We live in a consumer society, where clever marketing makes us believe that we need new products and technologies and more material goods. Products are designed and manufactured for mass markets/audiences around the world.

You should know...

☐ about the uses, advantages and disadvantages of built-in obsolescence and offshore marketing

☐ about cultural issues when designing products.

Key terms

built-in obsolescence   offshore manufacture

Which of the key terms best fits the description below?

A product that is designed to stop working after a certain period of time.

To check your answers, look at the glossary on page 173.

Multiple-choice questions

1. Which of the items would have been designed and manufactured with built-in obsolescence?

A A biro pen   **C** A DVD film

B A cake   **D** A pair of glasses

2. Offshore manufacturing is best described as:

A making products in China

B large manufacturing companies and industries relocating their businesses from one country to another

C taking advantage of cheap labour costs in India

D making products out at sea

Practice exam questions

1. Manufacturers make products with built-in obsolescence because:

   A they enjoy making things

   B they think consumers can afford new products

   C they continue to make money

   D they like to think that consumers are happy to keep buying new products.

   (1)

2. A UK-based manufacturing business is moving its company to China.

   (a) Describe **two** disadvantages to the local UK-based manufacturing industry of moving the business to China. (4)

   (b) Describe **one** environmental disadvantage of the manufacturing company moving its business to China. (2)

Manufacturers change the style of a product for fashion reasons.

Give *three* moral reasons against changing the style of a product for fashion reasons. (3 marks, 2008)

Note: The data and comments below relate to the first response given by the student in relation to the question above.

Student answer	Examiner comments	Build a better answer
■ *It makes things dearer* (0 marks)	This type of response is too general.	△ Prices go up due to consumer demand for new products and gadgets.
△ *People are made to feel poor if they do not have the newest products* (1 mark)	A good understanding about issues relating to peer pressure and the consequences if they do not keep up to date with new products.	△ Peer pressure can lead to bullying in order to maintain status within a group.
△ *Waste of materials* (1 mark)	The student's basic understanding is correct but the answer could give a little more detail.	△ More of the planet's valuable raw resources are used in making new materials.

Overall comment: This type of question only requires a basic level of response without any justification. The majority of students were able to give at least one moral reason against changing the style of a product for fashion reasons.

Give one reason why manufacturers make products with built-in obsolescence. (1 mark, 2004)

Student answer	Examiner comments	Build a better answer
■ *To make money* (0 marks)	This level of response is far too basic to achieve a mark.	△ In order to maintain profits and turnover.
■ *Keep fashion moving* (0 marks)	This response makes a wrong judgment about how fashion and trends move.	△ Keep up to date with fashion and trends.
● *suck in new consumers* (1 mark)	A correct response but the language is a little crude.	△ Create new markets.

Overall comment: On the whole this question was quite well answered, but it was clear that many students did not understand the term built-in obsolescence.

Your controlled assessment

Controlled assessment

In the controlled assessment task, you will demonstrate your skills, knowledge and understanding by designing and making a high-quality resistant materials technology (RMT) product. You can choose a linked activity where you design and make one product, or a separate activity, where you design one product and make another. You must select your task from a list provided by the exam board; this list is updated every two years, and will be provided to you by your teacher.

What is a resistant materials technology product?

A resistant materials product is a fully-functioning product that matches its specification. It must be manufactured to full size using resistant materials, such as those defined in Chapter 1.

Your assessment

The controlled assessment task is worth a total of 100 marks. The table on the next page shows how these marks are split between the different parts of the task and what is involved. It is recommended that the controlled assessment task take about 40 hours but this will not be recorded or monitored.

What are the controls in the controlled assessment?

Some parts of the controlled assessment task must be carried out under supervision in the classroom or workshop, and some can be done away from school. This affects where and how you work.

Preparation: You can do some research and preparatory work outside of the classroom without supervision.

Write-up and making: You will have about 40 hours to complete the design and make activities.

The write-up does not have to be done under supervision at all times, but the majority of the work you present for assessment must be carried out in the classroom or workshop. You may use ICT to process some of your write-up, as long as this work has been drafted in the classroom under supervision and that the final version matches the original draft.

Making activities should be done under supervised conditions in accordance with health and safety regulations too. This can prove helpful because your teachers are allowed to give you demonstrations of any new processes and techniques. Another thing to remember is that you are not allowed to take your practical work out of the classroom.

How the task will be assessed

Design activity (50 marks)		Make activity (50 marks)	
You will undertake a design activity covering the following three stages and eight assessment criteria:		You will undertake a make activity covering the following three stages and five assessment criteria:	
Stage 1 Investigate (15 marks)		**Stage 4 Plan (6 marks)**	
1.1	Analysing the brief (3 marks)	4.1	Production plan (6 marks)
1.2	Research (6 marks)		
1.3	Specification (6 marks)		
Stage 2 Design (20 marks)		**Stage 5 Make (38 marks)**	
2.1	Initial ideas (12 marks)	5.1	Quality of manufacture (24 marks)
2.2	Review (4 marks)	5.2	Quality of outcome (12 marks)
2.3	Communication (4 marks) evidenced throughout	5.3	Health and safety (2 marks) evidenced throughout
Stage 3 Develop (15 marks)		**Stage 6 Test and evaluate (6 marks)**	
3.1	Development (9 marks)	6.1	Testing and evaluation (6 marks)
3.2	Final design (6 marks)		

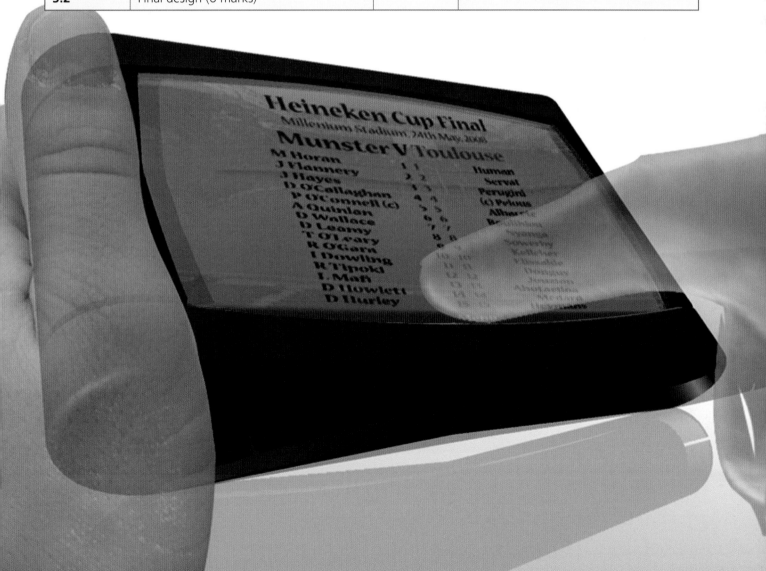

Chapter 9 Design
Introduction

In this section of your controlled assessment task you are expected to carry out a single design task. You will need to work in one of the tasks set by the exam board; your teacher will inform you about these since they are subject to change every few years. This task will allow you to show off your designing skills when you tackle a project, such as designing a desk-top organiser that is required to hold a collection of stationery items, including pens, pencils, paper clips and scissors.

Choosing your controlled assessment route

There are two ways to complete your controlled assessment task.

1 You can work through all six stages in order, beginning with Stage 1, which means you make what you design.
2 Alternatively, you can split the six stages into two parts: design activity (Stages 1–3) and make activity (Stages 4–6). If you choose to do this, the product you design as part of the design activity will not be the same as the product you make as part of the make activity. You might find this to your advantage if you find it difficult to sustain interest in a single project over a long period of time. Another advantage is that you can be a bit more adventurous with your design because you do not have to make what you design. An example of this is shown on page 127, which is a 'blue sky' handheld sports communication device. Alternatively you might choose to make something more challenging than you have designed.

Your assessment

The design activity is worth 50 marks, which is 50 per cent of the marks available for your controlled assessment task.

The design activity is split into stages and you will need to produce evidence in all three. These stages are:

1 Investigate
2 Design
3 Develop

Controls in the design activity

You must complete, under supervision, the write-up of your portfolio and the making of your product. The majority of work presented for assessment must be carried out in the classroom or workshop.

How the task will be assessed

Design activity (50 marks)				
You will undertake a design activity covering the following three stages and eight assessment criteria:				
Stage	**Assessment criterion**	**Marks**	**Suggested times**	**Suggested pages**
1. Investigate	1.1 Analysing the brief	3	1 hour	1
	1.2 Research	6	3 hours	2–3
	1.3 Speciification	6	1 hour	1
2. Design	2.1 Initial ideas	12	5–6 hours	2–3
	2.2 Review	4	1 hour	1
	2.3 Communication evidenced throughout	4	Evidenced throughout design and development stages	N/A
3. Develop	3.1 Development	9	5–6 hours	2–3
	3.2 Final design	6	1–2 hours	1–2
	TOTAL	50	17–20 hours	10–14

Feverpitch

'The live game, in your control'

Stage 1 Investigate (15 marks)

128

Objective

- **Carry out** a detailed analysis of the design brief.

Watch out!

Make sure that you include your design brief, clearly labelled, on the first page of your project folder.

Control

You need to carry out most of this section under the supervision of your teachers at school and not in your own time for homework away from school.

1.1 Analysing the brief

You should spend about 1 hour on this section. It is worth a total of 3 marks.

To get top marks (3) you will:
make sure that your design brief or task is clearly shown on the first page alongside your analysis
cover most design needs, exploring and clarifying them fully, when you analyse your design brief or task.

Your design brief should be a short statement outlining what you are going to design.

Your teacher might give you a design brief based on one of the themes set by Edexcel, or you could think of your own. If you do provide your own design brief, it must fit within the exam board's set themes. These themes are reviewed every few years; your teacher will tell you what they are when you start your controlled assessment task.

When you analyse your design brief, highlight all the areas you will need to consider when you start to gather your research and design ideas. The brief should also help you to understand the issues related to your chosen design task. You might need to consider the materials and components that you could use, the function and performance of your product, and any related sustainability issues.

Some other areas for further consideration are:

- form
- function
- user requirements
- materials and components
- performance requirements and
- sustainability.

Presenting your ideas

You need to think about how you are going to present your information in this section. One of the best methods is a mind map. This will allow you to show areas for investigation and research in a clear and precise fashion.

For each of the points above further detail is required which is relevant and specific to your own design task. In the area of user requirements, for example, you will need to consider what the user actually wants the product to do and how they interact with it, such as getting out the wine bottles without the risk of knocking the unit or rack over.

You also need to make sure that your design brief is clearly identified on your first page so that the moderator can see exactly what you are going to design.

Investigate – Analysing the brief

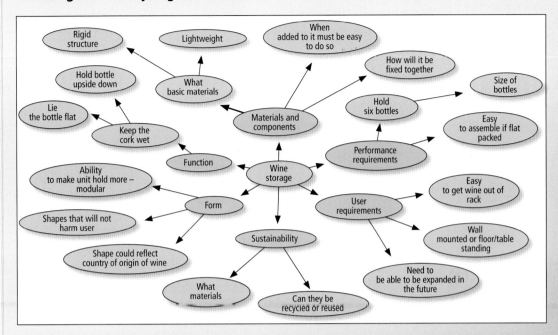

In the spider diagram on the left, I have tried to think about all the areas that I will need to consider as I start my project. The areas in blue are the key areas and I will need to try to find out some answers through my research to the questions posed.

Design needs – I need to investigate:
- The different size of wine bottles to hold
- How to access the bottles
- How they can be stored to keep the cork wet
- Where the wine comes from and what influences I might be able to use from this
- Sustainability – what will happen to the product once it reaches the end of its working life
- What products already exist that might be able to help inform my specification and design ideas

Design Brief
To design, make, test and evaluate a wine rack. It must be capable of holding 6 bottles and capable of being expanded in the future to hold more.

Moderator's comments

The student has used a spider diagram to highlight the main areas for consideration as they start to carry out their research. The key areas have been highlighted and, from them, others issues and areas have been raised. The design brief is very clear and succinct, saying exactly what they are going to do. The brief fits into the areas specified by the exam board in that this project will fit into the storage section. Other design needs have been raised and explained that will enable the student to be able to go on and carry out some specific, focused research.

The work of the student here is judged as:	
Needs improvement	
Average	
Good	✔
Very good	

Objectives

- **Carry out** and present selective and focused research that addresses the design needs of your product.

- **Analyse** the performance of a relevant existing product using product disassembly.

- **Apply** the findings from your product analysis and research to develop your own design specification.

ResultsPlus
Watch out!

Remember that your research must be selective and focused. Concentrate on your identified needs, and don't include unnecessary details.

1.2 Research

You should spend about 3 hours on this section. It is worth a total of 6 marks.

To get top marks (6) you will:
make sure that your research is focused and relevant to the design task
consider and explore performance, materials, components, processes, quality and sustainability issues
make good use of a product analysis exercise to help you understand the nature of the design task better and to write your specification.

Using your analysis as a starting point, carry out some research to help you develop a design specification. Your research must be focused and relevant. For example, if you were planning to design a wine bottle storage rack, you should look at the size of the bottles you want to store. This will help you when it comes to the design stage.

You should carry out a product disassembly task on a relevant existing product. This will help you to understand how the product performs, the materials and components involved, which processes have been used, and the relevant quality and sustainability issues.

You should use the work completed in the previous section as your starting point for this section. You should have identified key areas when you analysed your design brief, which you should now use as headings to focus your research.

It is important that you consider the sizes and dimensions of items that you might be holding or storing. When you start designing, you will need to use this information to make sure that your product will be big enough for its intended purpose. Here you can use photographs or drawings with dimensions annotated on them.

You also need to analyse an existing, similar product. Here you should consider:

- materials used – including sustainability-related issues
- dimensions
- how the product is assembled
- what packaging is used
- form
- performance requirements and quality.

Make good use of photographs when carrying out your product analysis. You could also use a table to record your observations, using the headings above.

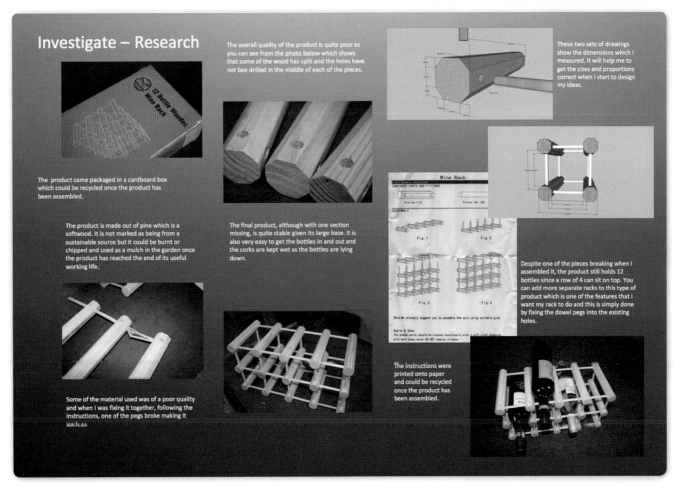

Investigate – Research

The product came packaged in a cardboard box which could be recycled once the product has been assembled.

The product is made out of pine which is a softwood. It is not marked as being from a sustainable source but it could be burnt or chipped and used as a mulch in the garden once the product has reached the end of its useful working life.

Some of the material used was of a poor quality and when I was fixing it together, following the instructions, one of the pegs broke making it useless.

The overall quality of the product is quite poor as you can see from the photo below which shows that some of the wood has split and the holes have not bee drilled in the middle of each of the pieces.

The final product, although with one section missing, is quite stable given its large base. It is also very easy to get the bottles in and out and the corks are kept wet as the bottles are lying down.

The instructions were printed onto paper and could be recycled once the product has been assembled.

These two sets of drawings show the dimensions which I measured. It will help me to get the sizes and proportions correct when I start to design my ideas.

Despite one of the pieces breaking when I assembled it, the product still holds 12 bottles since a row of 4 can sit on top. You can add more separate racks to this type of product which is one of the features that I want my rack to do and this is simply done by fixing the dowel pegs into the existing holes.

Moderator's comments

The student has carried out a product analysis on a wine rack. They have assembled it and taken some photographs to show the various stages. They have also made some objective comments on the quality of the materials, the packaging, the materials and some sustainability issues. They have suggested how they might be able to use some of this information when they start their design work, such as the dimensions and proportions of the materials used.

The work of the student here is judged as:	
Needs improvement	
Average	
Good	
Very good	✔

Objectives

- **Produce** realistic, technical and measurable specification points which address some issues of sustainability.

- **Use** the findings from your research to justify these specification points.

ResultsPlus
Watch out!

Use your research to justify each of the points listed in your specification.

1.3 Specification

You should spend about one hour on this section. It is worth a total of 6 marks.

To get top marks (6) you will:
make sure that the points listed in your specification are realistic, technical and measurable
include some issues related to the sustainability of your product
justify fully all specification points.

Your specification is very important because you will use it as a framework when you design your product. It should consist of a list of points, backed up by your research, detailing what your design should do and why: for example, 'The wine storage system should hold a minimum of six bottles because that is the average number of bottles that the people surveyed have at home.' You will also use your specification when reviewing your design ideas as they progress.

Your specification will be best laid out in a table, with each point clearly identified and followed by the justification for its inclusion.

Your specification should cover the following areas:

- technical points: for example, 'It must be tough and capable of withstanding sudden impacts so that it does not fall apart'.

- measurable points: for example, 'It should not weigh more than 15 kg. This is because similar products that I looked at weighed a maximum of 18 kg and, if it is too heavy, it will become too difficult for the user to move.'

- sustainability: for example, 'The timber used must be from an FSC approved source, as these materials will come from managed sources and will go some way to ensuring that we maintain our forests and potential supply chains.'

Point	Reason
List your specification headings in this column	Make sure that your points are fully justified and explained in this column

Investigate – Specification

Area	Point	Reason
Form: Why is the product shaped as it is?	The wine rack must : • have a stable base	• so that it does not fall over and allow the bottles to fall out and break
Function: What is the purpose of the product?	• hold six bottles of wine	• because from my survey questionnaire, that was the average number of bottles that potential users had at home
User requirements: What qualities make the product attractive to potential users?	• be capable of being extended in the future	• so that if you end up needing to store more than 6 bottles in the future, you can "add on" extra storage space
Performance requirements: What are the technical considerations that must be achieved in the product?	• allow easy access to the bottles • store the bottles so as to keep the corks wet	• so that it is easy to get them in without dropping them • so that the cork does not become dry, shrink and let air in which will spoil the taste of the wine
Material and component requirements: How should the material and components perform within the product?	• produce a rigid structure	• so that the bottles will be held in a structure that will not collapse and hold them firm
Sustainability: How does the design allow for environmental considerations?	• be made from materials that can be recycled	• so that once the product reaches the end of its useful working life it can be recycled rather than be thrown away into a landfill site

The photo on the left is the wine rack that I carried out my product analysis on. It shows two different ways that the bottles can sit on it but it also shows the holes that are left exposed in the top row so that another complete rack can be added onto it, making it capable of holding even more bottles in the future.

Moderator's comments

The student has produced a simple specification for their wine rack. They have used some of their findings from their research to justify and support their points such as the number or bottles to hold. They have also included some of their product analysis work to show how the wine bottles could be held and the whole rack expanded in the future if needs be.

However, the reference to sustainability is a little bit vague.

The work of the student here is judged as:	
Needs improvement	
Average	
Good	✔
Very good	

Stage 2 Design (20 marks)

Objectives

- **Present** alternative initial design ideas that are realistic, workable and detailed.

- **Demonstrate** your understanding of materials, processes and techniques.

- **Apply** your research findings to your design work.

- **Address** specification points through annotation.

Make sure that you address each of the specification points fully and that your annotation relates to them clearly.

Control

You need to carry out most of this section under the supervision of your teachers at school and not as homework in your own time away from school.

It is better to produce fewer designs in greater detail than to have a lot that do not reach the higher levels of response.

2.1 Initial ideas

You should spend between 5 and 6 hours on this section. It is worth a total of 12 marks.

To get top marks (12) you will:
make sure that your initial design ideas are realistic, workable and detailed
demonstrate a clear understanding of materials, processes and techniques
use annotation and research information to address all key specification points.

This is your opportunity to demonstrate your creativity and flair by producing a wide range of different initial design ideas in response to your original design need. Use a range of communication techniques to present your initial ideas. You could include 3D sketches, 2D views, exploded sections and ICT where appropriate. Each design solution should be complete and should address all the specification points. Use annotation to support your graphical work with details about materials, processes and techniques that could be used in manufacturing the product. You must use your research findings to support your design work; remember to include these in the annotation.

At this stage, when you start to produce your initial ideas in response to your design brief and specification, you need to concentrate on producing high-quality solutions that include a good level of detail. This is far better than producing lots of ideas that lack detail and do not reach the higher levels of response as set out in the assessment criteria. For this reason, your individual design ideas should be well annotated to show what materials, processes and techniques might be used if the idea were to be made.

You also need to make sure that your annotation addresses the specification points detailed in the earlier section. Use your specification points too as a guide when coming up with design ideas, to make sure that your solutions are workable and realistic. For example, if your specification says that the wine rack must hold a maximum of six bottles, do not present solutions that hold more than six bottles.

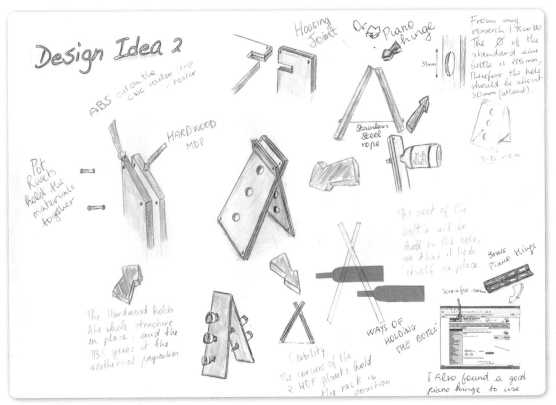

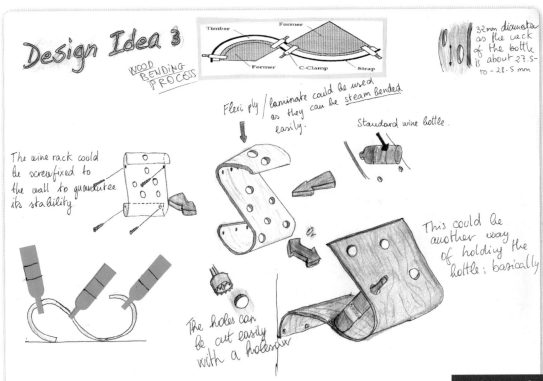

Moderator's comments

This student has produced imaginative design ideas that are realistic and workable, with some good detail. They have shown a good understanding of a range of materials and processes that could be used in their manufacture. Ideas are well annotated and address many of the specification points.

The work of the student here is judged as:	
Needs improvement	
Average	
Good	
Very good	✔

Objectives

- **Present** objective evaluative comments against the original specification criteria.

- **Evaluate** your initial design ideas by taking into account user group feedback and issues of sustainability.

ResultsPlus
Watch out!

Review all your specification points and make objective comments about them. Don't just use a simple 'yes/no' score sheet tick boxes, marks out of 10 or '1–5' scoring system to work out which design scores the highest mark.

2.2 Review

You should spend about one hour on this section. It is worth a total of 4 marks.

To get top marks (4) you will:
evaluate your initial design ideas objectively
take into account user-group feedback and issues of sustainability.

You need to decide which of your ideas it would be best to develop further. To do this, you must review and evaluate your initial design ideas against the original specification criteria. Your comments must be objective rather than subjective: this means making supported statements about your designs that are fully explained such as:

- **This design is favoured because when I asked potential users they felt it would be the most stable design, as it had a large base area.**

Rather than subjective statements such as:

- **I like it very much and it is a nice colour.**

It is always a good idea to ask the people who might use your product what they think of your designs. Users' opinions will help you to understand how you could make your designs even better.

Your work for the review section is best laid out in tables like these.

Specification point	Idea 1	Idea 2	Idea 3

	User-group feedback
Idea 1	
Idea 2	

The comments that you and potential user groups make should be recorded in the tables, and the comments should be objective.

Avoid using tick/cross boxes to record whether the idea is a good one or not; similarly avoid giving ideas a score and adding up to see which ones score the highest mark. Do try to make sure that you address most of the initial specification points.

Review

	Design Idea 1	Design Idea 2	Design Idea 3
Form	• The design is stable and has a wide enough base area to suit a regular floor. • It is aesthetically pleasing because of the properties of the wood. • All the bottles are visible.	• It is stable as the edges of the structure put friction on the floor. • It is aesthetically pleasing as the combination of ABS/wood suits it. • The bottles are mostly visible as we can see the labels/forms.	• This design is attached to a wall, which makes it very stable and safe. • The curved wood gives it an amazing effect. • The bottles are put in holes which make them visible.
Function	• The structure can hold up to 12 wine bottles; however it is not able to hold any glasses. • It is stable as it has a wide area.	• It holds up to 6 bottles and no glasses, so it fails my spec. • It is stable as the structure has a very wide base.	• It will be able to hold up to 10 wine bottles but no glasses. • It is stable as it is attached to a wall.
User Requirements	• It will be done at a high standard, which will mean that I need to put a lot of time in. • It will be able to hold 2 different-sized bottles as the strips the bottles lie on will have different-sized spaces.	• As the previous one it will be at a high standard if produced. • It will hold at least two different-sized bottles because of the diameter of the holes the bottlenecks fit into, to suit the wine bottles.	• As the previous two its aesthetic properties will depend on the time put in. • The way it will hold different-sized bottles will be the same as idea 2.
Safety	• It will be stable because of the 4 feet and the wide base. • It will hold the bottles safely as the bottles will lie with almost their full horizontal length on the strips of wood.	• It will be stable as the edges of the structure will put friction on the floor and the wide base will help. • It will hold all the bottles safely due to the friction of the bottlenecks on the holes.	• It will be very stable, as it will be attached to a wall by rivets. • The bottles will be held in place by the same principle as idea 2, it will be very safe and the bottles will be untouched.
Budgetary Constraints	• It won't overrun the £150 boundary as it is mainly made out of wood and its only metal parts are rivets, which will not cost much. This succeeds in achieving my specification point.	• This one should cost about £100 or more. It is more expensive than the previous one as there are more materials used in this case e.g. acrylic. So it achieves my specification point.	• This one will touch the budgetary boundary as it involves some processes and it will use some other features which will make it more expensive. It might not be successful cost-wise.
Materials and Components	• It follows my specification point as it is made entirely out of hardwood, has mortise and tenon joints, and the feet are attached with rivets from the bottom of the structure.	• It follows my specification points as it is made out of 2 materials mentioned in my research: acrylic and MDF. A housing joint or a brass piano hinge will hold the two main parts together.	• This one might be made out of flexi ply or laminate, as they can be both bent. It will use rivets to be attached to the wall. Therefore it follows all my specification points.

Moderator's comments

The student has produced a table that contains some good evaluative, subjective comments relating to their initial design ideas. They have made reference to many original specification points but they have not considered the views of the user group or sustainability, so they will not be able to score full marks in this section.

The work of the student here is judged as:	
Needs improvement	
Average	
Good	✔
Very good	

138

Objectives

- **Use** a range of communication techniques and media, including ICT and, where appropriate, CAD.

- **Use** communication techniques with precision and accuracy.

Watch out!

Your whole design folder should show evidence of communication, so do make sure that you include a range of techniques, including ICT where appropriate.

2.3 Communication

This section is demonstrated throughout your design and development work. It is worth a total of 4 marks.

To get top marks (4) you will:
use a range of communication techniques precisely and accurately
include ICT and CAD where appropriate.

Your communication skills will be in evidence throughout your design folder but you should try to show a range of appropriate techniques in the design and development stage. Use 2D and 3D drawings to convey your ideas. Exploded views show detail and how separate components fit together. Word processing, digital photography, graphs and tables are all appropriate uses of ICT and may be used in the investigation section. CAD can be used as a modelling tool in the development stages. You must also produce a final design drawing that includes sufficient technical detail about materials and component parts to enable the product to be manufactured.

You will need to show off your communication throughout your work, but the most obvious and natural sections in which to do this are the design and development sections.

Here you should try to use as many of the following techniques as you can:

- 3D sketching
- 2D sketching
- computer modelling and CAD packages
- section drawings
- exploded views to show how things fit together
- orthographic drawings to show dimensions.

Use these techniques with precision and accuracy to demonstrate your design ideas and thinking clearly.

Moderator's comments

The student has used a wide range of communication techniques such as 2D and 3D sketching, exploded views, word processing, digital photography, PowerPoint and CAD. The dimensions on the CAD drawings help to show how big the analysed existing product is. They have also included a scanned set of written instructions to analyse. All of these techniques have been used with precision and accuracy, demonstrating that their ideas have been thought through clearly.

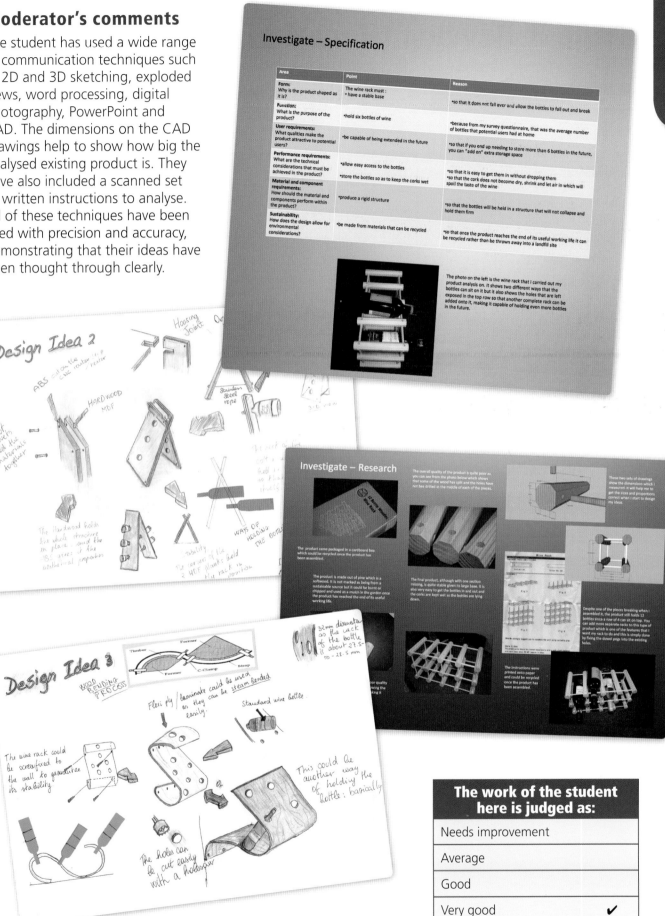

The work of the student here is judged as:	
Needs improvement	
Average	
Good	
Very good	✔

Stage 3 Develop (9 marks)

Objectives

- **Develop** your initial design ideas into a single, final design proposal.

- **Use** scale modelling, including 2D/3D modelling in traditional materials and/or 3D computer simulations.

- **Evaluate** your ideas against the original design specification.

Results Plus
Watch out!

Your final design proposal must be significantly different from and better than any other of your ideas.

Control

You need to carry out this section under the supervision of your teachers at school and not as homework in your own time away from school.

3.1 Development

You should spend about 5–6 hours on this section. It is worth a total of 9 marks.

To get top marks (9) you will:
make sure that your final design proposal is significantly different from any other design idea
make a model to test important aspects of the design idea
produce an evaluation of the design proposal against the original design specification.

This section is about developing your initial design ideas into a single and improved final design proposal. You should bring different aspects and features of your initial ideas together in one, final design proposal that is significantly different from and better than any single initial design idea. Combine the best parts of your ideas, before modelling aspects to test and further improve if necessary. Traditional modelling using cardboard, foam and conventional materials is acceptable, or 3D or computer modelling and simulations. Use any modelling constructively for testing and as a basis for further improvements. Your final design proposal should be evaluated against the original design specification to ensure that it meets all aspects required.

Any modelling, be it cardboard modelling or 3D computer modelling, must be done for a specific purpose. Modelling is not a hoop that has to be jumped through and done just for the sake of doing it.

You should use modelling to test aspects or features of an idea, and evaluate them against the relevant design specification points. The aspects and features modelled should also be evaluated by user groups, so that you can take their feedback into account and make modifications before presenting your final design proposal.

You should photograph any modelling so that you can include it in your design folder, ready for assessment by your teachers.

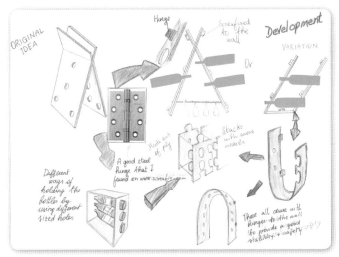

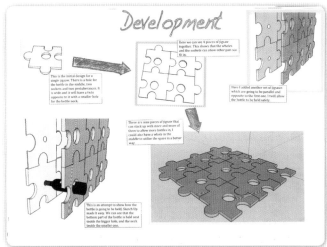

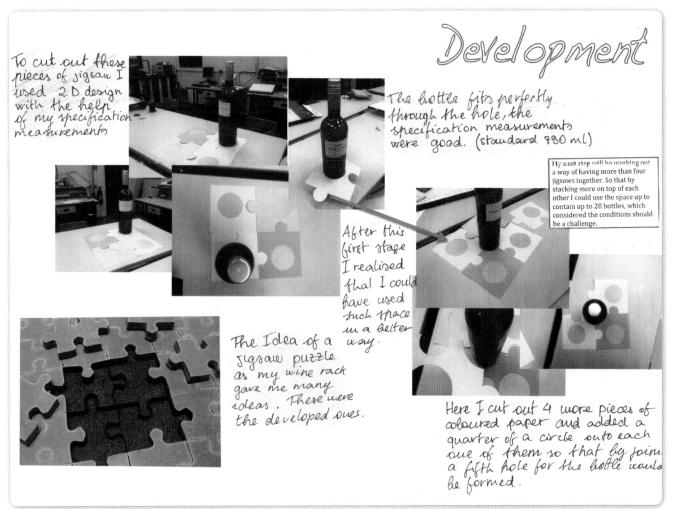

To cut out these pieces of jigsaw I used 2D design with the help of my specification measurements

The bottle fits perfectly through the hole, the specification measurements were good. (standard 780 ml)

My next step will be working out a way of having more than four jigsaws together. So that by stacking more on top of each other I could use the space up to contain up to 20 bottles, which considered the conditions should be a challenge.

After this first stage I realised that I could have used such space in a better way.

The Idea of a jigsaw puzzle as my wine rack gave me many ideas. These were the developed ones.

Here I cut out 4 more pieces of coloured paper and added a quarter of a circle onto each one of them so that by joining a fifth hole for the bottle would be formed.

Moderator's comments

The student has developed their work through the use of some external sources. The final solution as it stands here is significantly different and improved in comparison with any of their other design ideas. They have used some 3D computer drawings to show what it will look like, but there is little evidence on these pages to show any user-group feedback. Excellent use has been made of some paper modelling to gauge overall proportions.

The work of the student here is judged as:	
Needs improvement	
Average	
Good	
Very good	✔

142

3.2 Final design

You should spend about 1–2 hours on this section. It is worth a total of 6 marks.

To get top marks (6) you will:
make sure that your final drawings include technical details of materials and/or component parts, processes and techniques.

You should prepare and present your final design proposal in the form of a set of drawings. These drawings should include technical details relating to the sizes of component parts and details of the materials, processes and techniques to be used in manufacturing the product.

The drawings and annotation should be detailed enough to allow someone else to make the product. Drawings can be in 2D or 3D. You might include some exploded views to show how component parts are to be assembled and fitted together. It is essential to give details of joints and dimensions.

Your work in this section is best presented as a series of drawings, possibly with some details in a table. Your drawings should show your final design proposal – what it is that you would go on to make – or should act as a presentation to a group of possible users of the finished design.

Your drawings will probably comprise:

- a collection of 3D views to show what it looks like and, if relevant, how it all fits together or works;

- 2D drawings should show specific details, such as overall or individual component part dimensions. You should give each part a part name or number.

You should use a table should to record details about individual parts, including the materials and processes used to make the part and what finish is going to be applied.

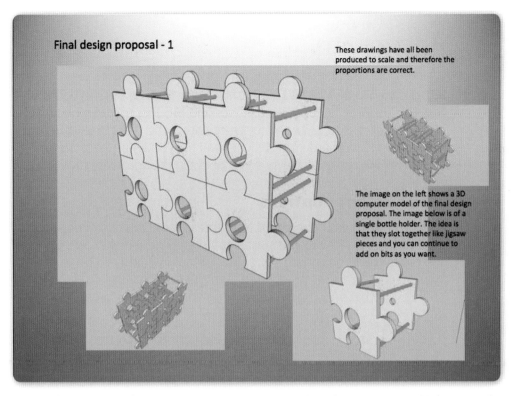

Final design proposal - 1

These drawings have all been produced to scale and therefore the proportions are correct.

The image on the left shows a 3D computer model of the final design proposal. The image below is of a single bottle holder. The idea is that they slot together like jigsaw pieces and you can continue to add on bits as you want.

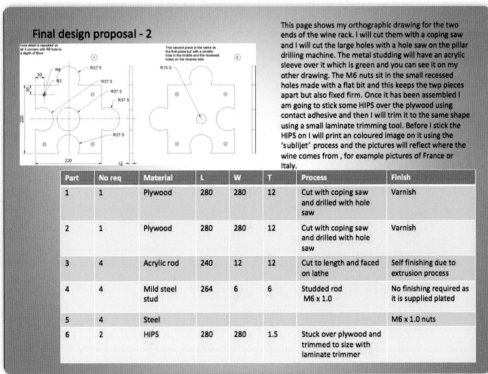

Final design proposal - 2

This page shows my orthographic drawing for the two ends of the wine rack. I will cut them with a coping saw and I will cut the large holes with a hole saw on the pillar drilling machine. The metal studding will have an acrylic sleeve over it which is green and you can see it on my other drawing. The M6 nuts sit in the small recessed holes made with a flat bit and this keeps the twp pieces apart but also fixed firm. Once it has been assembled I am going to stick some HIPS over the plywood using contact adhesive and then I will trim it to the same shape using a small laminate trimming tool. Before I stick the HIPS on I will print an coloured image on it using the 'sublijet' process and the pictures will reflect where the wine comes from , for example pictures of France or Italy.

Part	No req	Material	L	W	T	Process	Finish
1	1	Plywood	280	280	12	Cut with coping saw and drilled with hole saw	Varnish
2	1	Plywood	280	280	12	Cut with coping saw and drilled with hole saw	Varnish
3	4	Acrylic rod	240	12	12	Cut to length and faced on lathe	Self finishing due to extrusion process
4	4	Mild steel stud	264	6	6	Studded rod M6 x 1.0	No finishing required as it is supplied plated
5	4	Steel					M6 x 1.0 nuts
6	2	HIPS	280	280	1.5	Stuck over plywood and trimmed to size with laminate trimmer	

Moderator's comments

The student has produced two separate pages for this section. One page shows some 3D computer-generated images and the second shows an orthographic drawing of the two separate end pieces required. A table on the second page includes some detail about the materials required and the processes and finishes involved.

The work of the student here is judged as:	
Needs improvement	
Average	
Good	
Very good	✔

Chapter 10 Make
Introduction

In this section you will be able to show your making skills as you are required to manufacture a single, working product that can be tested and evaluated. You will need to select one of the tasks that have been set by the exam board. Your teacher will tell you what these areas are since they are subject to review and possible change every few years.

Make activity options

At this stage of your controlled assessment task, you can choose to stop the design activity and start a completely new make task, or make the product you designed. You should discuss whether to do this with your teacher. It makes no difference to the marks or the assessment criteria whether you choose to carry on with the same product or start again.

If you are going to follow the same task right the way through and make what you have designed, you will take your final design proposal forward into the make activity.

If you have decided to start a new task for the make activity, you will need a detailed set of working drawings. Your teacher might provide these or you might find them yourself, from a magazine or catalogue, for example. You must make sure that whatever product you choose to make allows you to demonstrate a wide range of skills and processes.

Your assessment

The make activity is worth 50 marks, which is 50 per cent of the marks available for your controlled assessment task.

The make activity is split into stages and you need to produce evidence in all three. The three stages are:

1 Plan
2 Make
3 Investigate.

Controls in the make activity

The make activity should be carried out under supervised conditions in accordance with health and safety regulations. Your teacher can give you demonstrations of all new processes and techniques. You are not allowed to take your practical work out of the classroom.

When you have finished your teacher will mark your project before it is sent to the moderator.

How the task will be assessed

Make activity (50 marks)

You will undertake a make activity covering the following three stages and five assessment criteria:

Stage	Assessment criterion	Marks	Suggested times	Suggested pages
4. Plan	4.1 Production plan	6	1–2 hours	1
5. Make	5.1 Quality of manufacture	24	16 hours practical	1–2
	5.2 Quality of outcome	12		1
	5.3 Health and safety evidenced throughout	2	Evidenced throughout make stage	N/A
6. Test and evaluate	6.1 Testing and evaluation	6	1–2 hours	1–2
	TOTAL	50	18–20 hours	4–6

These pictures show the manufacturing process in each stage of production. The numbers refer to the production plan on a previous page.

Photos to show manufacturing process

1/4/6. Mark out oak planks for legs and frame to be made

2/9. Cut out legs and frame

3. Planing

5. Drill our

7. Cut out tenons

8. Chisel tenons and mortise

10/13/19. Cut steel rod for handle

11/16. Drill and tap each end

22. Bending handle to 90° on either side

20. Cutting grill to size

21. Weld grill to frame

21. CNC router to cut lattice base

Stage 4 Plan (6 marks)

Objectives

- **Produce** a detailed production plan that considers the stages of manufacture for your product.

- **Identify** and describe the stages during making where specific quality control procedures should take place.

Watch out!

Plan how you are going to make your product as a one-off product, not how it would be made in volume using industrial and commercial processes. Remember, you are making a plan, not recording processes as you complete them.

Control

You need to carry out most of this section under the supervision of your teachers at school and not in your own time for homework away from school.

4.1 Production plan

You should spend about 1–2 hours on this section. It is worth a total of 6 marks.

To get top marks (6) you will:
produce a flowchart showing all the main stages of manufacture
indicate where quality control (QC) checks will be made
show the making stages set against time to show how you will meet deadlines.

You must plan the manufacture of your product carefully, which means that you must produce a production plan to show the various stages of manufacture. This plan could be in the form of a flowchart showing the stages of manufacture in the correct sequence and how long you think each stage will take. Highlight the stages where quality control (QC) will take place. Give specific details of what is being checked and how; do not just use simple generic statements such as 'check here for quality'. For example, when any material is being cut to size, you should know the size in advance. Make sure the size is correct by measuring it with a ruler or micrometer for more accurate measurements.

Your production plan should show all the stages of manufacture for a one-off product, rather than showing how it would be made commercially in batches, or even in high volume. It is also important that it is a proper plan, done in advance, rather than being completed retrospectively, or as a photographic diary as it is being made.

Your quality control (QC) checks should be specific, appropriate and realistic and not just for the sake of doing them.

You should also consider carefully the time allocated for each stage, again as a realistic effort. You can add notes if the time allocated is either too much or not enough, but it is a good idea to say why the timings have gone wrong.

This is a step-by-step guide to show the production process for making my product. With details on equipment, timing, quality control and safety.

Production Plan

		Process	Equipment	Time of processes	Quality control	Safety of self	Safety of others
1	Wood frame	Mark out oak planks for legs and frame to be made	Rule, pencil, set square	20mins	Check measurements carefully	Apron, safety shoes, read instructions on equipment used, tie long hair back,	
2		Cut out legs and frame	Circular saw	40mins	Follow lines accurately	Apron, safety shoes,	Ventilation and guard in place
3		Planing	Planing machine	20mins	If plane one piece, plane all of them	Safety goggles to prevent damage to eyes, apron,	Ventilation, make sure area closed off, make sure stop buttons work.
4		Measure mortise on legs	Mortise gauge, rule	40min	Check measurements with rule before each piece	Apron, safety shoes,	
5		Drill out mortises	Mortise machine	20mins	Ensure drill is between the marks	Safety goggles, Apron,	Ventilation, make sure area closed off, make sure stop buttons work.
6		Measure tenons to size	Mortise gauge, rule, set square	40mins	Check with rule	Apron to prevent damage to clothing.	
7		Cut out tenons	Tenon saw, wood vice	60mins	Follow line accurately, cut to waste side of line	Apron,	
8		Chisel tenons and mortise to fit	Chisel	40min	Use a sharp chisel, take small sections off	Apron,	Make sure there are no potential slipping hazards nearby.
9		Rout edges for base	Circular saw	40min	Cut to the waste side of the line.	Safety goggles, apron,	Ventilation, make sure area closed off, make sure stop buttons work.
10	Metal handles	Cut steel rod for handle	Rule, engineer blue, scribe, hack saw	20mins	Check measurements.	Apron,	
11		Drill and tap each end	Metal lathe, using a 4.8mm drill for a M6 tap set, tap, tap wrench, lubricant	40mins	Use centre punch on lathe to centre.	Apron, safety goggles,	Close off area around lathe with markings or a barrier.
12		Bending handle to 90° on either side	Blow torch, bench clamp, large hammer, protective gloves	40mins	Place rod in clamp in the same place each time and bend to the same angle	Apron, heat protective gloves, Safety goggles	When carrying hot rod to clamp be careful of others. Mark on hot surfaces with chalk.
13	Grill Supports	Cutting mild steel rod to length and marking out	Hack saw	60mins	Check measurements before cutting, apply tolerances.	Apron,	
14		Milling one side flat, and drilling holes for supports	Milling machine	40mins	Keeping bar in place,	Apron, safety goggles,	
15		Faced off each end, drill pilot hole and drill end (5mm drill)	Metal lathe, centre punch, 5mm drill	20mins	Check correct size of drill. Counter sink centre	Apron, safety goggles,	Close off area around lathe with markings or a barrier.
16		Tap end with M6 tap set.	M6 tap, tap wrench, lubricant	40mins	Make sure tread is not drunken	Apron,	
17		Supports for grill cut, faced off on metal lathe	Metal lathe	60mins	Check speeds on lathe.	Apron, safety goggles,	Close off area around lathe with markings or a barrier.
18		Turn down for turning for thread	Metal lathe, die, die stock, lubricant	40mins	Make sure thread is not drunken.	Apron, safety goggles,	Close off area around lathe with markings or a barrier.
19	Grill	Cut mild steel rod to length	Hack saw, engineers blue, scribe, rule, bastard file	20mins	Check measurement before cutting	Apron,	
20		Cut grill to correct size	Snippers	20mins	Check measurement before cutting	Apron,	
21		Arc Weld grill/spot weld	Welder, mask, heat protective gloves	40mins	Check right angles	Apron, welding mask, heat protective gloves,	Protective cover around welder, verbally warn people that you are about to start welding.
22	Base	Cut out on CNC router. (lattice structure) and glued together	CNC router, Tool path software.	60mins	Check measurements and add tolerances		Keep area around machine closed off.

Quality Control- Involves identifying specific areas where errors could arise in the manufacturing of the product and setting up controls which will limit the possibility of these errors occurring. Ensuring these controls are met is importan in order to achieve a high quality product that will satisfy customers by supplying a product without defects.

Moderator's comments

This page shows a very good level of detail about the processes involved and the time allocated to each stage. There is a good order to the whole plan, although some of the quality control checks could have a little bit more detail about what measurements are to be checked and how. A good level of safety awareness is also recorded.

The work of the student here is judged as:

Needs improvement	
Average	
Good	
Very good	✔

148

Stage 5 Make (38 marks)

Objectives

- **Attempt** a challenging making task involving the manufacture of several different components using a range of materials, equipment, techniques and processes.

- **Select** tools, equipment and processes, including CAD/CAM where appropriate, for specific uses.

- **Demonstrate** a detailed understanding of the working properties of materials selected for a specific use.

- **Demonstrate** a wide range of making skills with precision and accuracy.

ResultsPlus
Watch out!

To ensure that you have plenty of evidence of using different tools, equipment and processes, avoid using more than 50 per cent CAD/CAM in the manufacture of your product.

Control

You need to carry out this section under the supervision of your teachers at school. You are not allowed to take your practical work home to complete it, as Edexcel must be sure that the work is your own.

5.1 Quality of manufacture

This section ties in with the following section relating to the quality of outcome. Together, 16 hours are allocated to the manufacturing stages. This section is worth a total of 24 marks.

To get top marks (24) you will:
select the correct tools, equipment and processes for the job with minimal guidance
show an appropriate understanding of the working properties of materials
use the tools to make several different component parts
demonstrate high-level making skills with precision and accuracy
use CAD/CAM where appropriate.

In this section you need to show your ability to use a range of tools, equipment and processes to make a quality product involving several components and different materials. You must make the best product you can.

The evidence for the marks in this section will be a written and photographic record of the stages of manufacture showing all the relevant processes and steps in detail. This step-by-step photographic diary should document all the tools, equipment and processes you have used. Annotate photographs as fully as possible to show all the decisions that you made during the manufacturing stages. For example, you should explain why you chose one method rather than another, and briefly describe the problems and any difficulties you encountered during that stage.

You will need to produce evidence for this section, as it is not acceptable to be awarded the marks simply because you have made a product. A selection of photographs will be the best way to show the tools, processes and techniques you have used; these photographs should be laid out neatly in order, to show the steps and sequences involved.

You should also include brief details of the properties and working nature of the materials used: for example, 'I used beech, a hardwood, because it is hard and tough, which means it will withstand the knocks and bumps it will be subjected to. It also finishes well and fits in with other beech products in the kitchen.'

However, be sure in the manufacture that you do rely too much on CAD/CAM. CAD/CAM should be used where appropriate, and certainly no more than 50 per cent of your manufacture should be carried out using CAM. Avoid making your whole product using laser cutters and CNC machinery.

These pictures show the manufacturing process in each stage of production.
The numbers refer to the production plan on a previous page.

Photos to show manufacturing process

1/4/6. Mark out oak planks for legs and frame to be made

2/9. Cut out legs and frame

3. Planing

5. Drill our mortises

7. Cut out tenons

8. Chisel tenons and mortise

10/13/19. Cut steel rod for handle

11/16. Drill and tap each end

22. Bending handle to 90° on either side

14. Milling

15/17/18. Metal lathe

20. Cutting grill to size

21. Weld grill to frame

21. CNC router to cut lattice base

Product assembled without base or grill supports.

Moderator's comments

The student has manufactured a high-quality product, which is clearly shown in the photographs. CAD/CAM has been used appropriately but does not exceed the 50 per cent rule. A wide range of tools and processes has been used with precision and accuracy, which again can be seen in the photographs. The production plan provides evidence of tools and processes selected. The task was very challenging.

The work of the student here is judged as:	
Needs improvement	
Average	
Good	
Very good	✔

Objectives

- **Produce** high-quality components that are accurately assembled and well finished, resulting in a high-quality product overall.

- **Produce** a completed product that is fully functional.

ResultsPlus
Watch out!

Your product must be complete and functional.

ResultsPlus
Watch out!

Health and safety are assessed in this section. You will gain two marks if you can demonstrate a high level of safety awareness throughout all aspects of manufacture. Your teacher will award the marks for health and safety based on their observations of your work during the manufacturing stages. You do not need to record risk assessments for your making activities, but you should always be thinking about your own safety and the safety of others around you.

5.2 Quality of outcome

This section ties in with the previous section relating to the quality of manufacture, and is part of the 16 hours allocated to the manufacturing stages. It is worth a total of 12 marks.

To get top marks (12) you will:
make sure that your product is of a high quality and has been accurately finished and assembled
produce a fully functional product.

Your final product must be identical to your final design proposal or your working drawings, depending on whether you are making the product you designed in the first part of the controlled assessment task or a different product.

You can use close-up photographs of component parts to show that they are of high quality and well finished. You can also take photographs as you assemble any component parts that might not be visible later on in a photograph of the final product.

You must complete your product so that you can show how it functions; some photographs of the completed product being used or tested will help to provide evidence of this.

Photographs are the best media to use to show the quality of outcome. Photographs of the product in situation are especially helpful, as are some photographs of the product being used and tested. If your product has to hold or store items, some of your photographs should show the product actually holding the items. However your main priority is to produce a fully functional product that is fit for purpose.

Moderator's comments

The student has produced a high-quality piece of practical work, which, when compared with the final design proposal, bears a very good likeness. The student has included in their final design proposal some photographs of the final product, and to demonstrate its fitness for purpose they have also included some photographs of the product being used.

The work of the student here is judged as:	
Needs improvement	
Average	
Good	
Very good	✔

Stage 6 Test and evaluate (6 marks)

152

Objectives

- **Devise** and **carry out** a range of suitable tests to check the performance and/or quality of the final product.

- **Evaluate** your final product objectively with reference to specification points, user group feedback and sustainability issues.

ResultsPlus
Watch out!

In this section you will be assessed on quality of written communication (QWC). Organise your information clearly and coherently, and use specialist vocabulary when appropriate.

Control

Most of this section needs to be completed under the supervision of your teachers at school. If you need to carry out interviews and testing, this can be done away from school.

6.1 Testing and evaluation

You should spend about 1–2 hours on this section. It is worth a total of 6 marks.

To get top marks (6) you will:
carry out a range of tests to check the performance and/or quality of the product with justification
put together a full, objective evaluation, including user-group feedback and consideration of sustainability issues.

You need to carry out a range of tests to check the performance of your final product. These might include using the product for its intended purpose and taking some photographs of it being used. You should also try to get the opinions and comments of some potential users to see what they think.

You must also evaluate the final product against the original design specification by making objective comments about how it meets or fails to meet the specification points. These comments must explain fully how and why the product meets or fails to meet the specification. Simple comments such as 'Yes, my product meets the specification point' are meaningless and will not score any marks. You must also describe how your product meets the sustainability issues raised in your initial specification. However, if you have completed a separate making and designing activity, you must test and evaluate your product against a set of of relevant manufacturing criteria. These criteria must contain some measurable criteria so that effective testing and evaluation can be carried out.

This is often the section students do poorly as many run out of time at the end of their projects. Make sure that you plan your time wisely and leave enough time to complete this section.

Your work here will probably best be presented in a table, with photographic evidence used to support your testing. Your table should have one column for your original specification points and a second column for your results and comments.

It is also very important that user groups use and test your product and provide some feedback.

Sustainability is a key area for consideration. Here are some useful headings to help you in this area.

- **Raw materials** What impact have they had on the environment? Could you have used fewer materials? Could you have used recycled materials?

- **Manufacture** Were the processes energy-efficient? Could the amount of waste be minimised?

- **Use** Will your product last a long time? Can it be repaired if it breaks?

- **End of life** Can your product be recycled or re-used?

Below is a copy of the questions I asked of the client along with his responses.

1. Do you think that the over all design of the product suits the area of the house?
   Yes the design is a very strong and robust solution. The range of materials used is good as the external area of the house has wood, metal and stone, two of which of these materials is used in the BBQ.

2. Do you think that the materials that the product is made from are suitable and aesthetically pleasing?
   I think that the materials used are a good compromise. The oak has been very well finished and the stainless steel is not going to rust when left outside, even in the open air.

3. Do you feel the size is suitable to cook food for up to 20 people?
   The BBQ is a good size and as we have tested it today it is very clear that the size is more than adequate for cooking for 20 people on.

4. Is the product suitable for cooking delicious food?
   As we have seen from the testing today and from the reaction of the boys in the house, the BBQ is more than suitable. The food was very good and the fact that we also used a probe to measure the temperature of the food, it was even more realistic.

5. Is the height and depth appropriate for the user?
   The size is very good, especially the height. I did not get any backache or pain whilst stood cooking at the BBQ for nearly an hour. The depth was also good and I did not have to stretch too far to be able to reach the back of the BBQ

6. Do you think there is enough storage space for equipment and food?
   There was a good amount of space for both food and equipment. I was able to cook the food and store it on the grill area without it getting cold. The rack below the table provides a good space for storing plates and tools etc. and it came in very useful when cooking.

7. What improvements do you think should be made?
   A lid would be useful to put over the unit once finished with a d a better method of being able to remove the burnt coals and dripping fat from the tray would be very useful.

To: ○ Barry Lambert
Cc:
Subject: Client comments on product testing
Attachments: none

Calibri

Dear Mr Lambert
Could we please arrange a time for us to meet to discuss your comments on the testing of the product. I am grateful that you took time to supervise the barbecue lunch today. It seemed to be a very successful occasion.
Look forward to hearing from you.

Will Ripley

A copy of the invitation to test my product. After the event they will given the questionnaire above to complete.

Conclusion: The testing of the barbeque was a successful social event, so much so that the pupils would like to be entertained using the barbeque several times over the summer, particularly when the exams have finished. Many year groups were present and it gave a good opportunity fro interaction between them. Pupils who weren't there have commented that they wished they had not missed it and would like to be present next time. The impromptu nature of the product being in place means that it can be used frequent with out too much forward planning. The students like the casualness of this idea rather than involving college catering. It also means that they can take part in cooking the food themselves.

Copy of email to client:

To all members of Southwood
There will be a barbecue lunch at 13.00hrs in the garden on Tuesday 5th May. You are welcome to come along and sample the food cooked on the barbecue that I have produced.
Looking forward to seeing you there.
Will Ripley

Interview with other potential users / customers.

Are you pleased with the finished product?

Do you find it aesthetically pleasing?

Are you happy with the size of the barbeque and do you think it is suitable for its need?

Do you feel that the product suits the style of the veranda area and garden?

Do you think it will be easy to move into the set storage over winter?

Are there any changes you would like to see?

Do you feel that it is value for money and would you be prepared to pay the full cost?

Testing and Evaluating

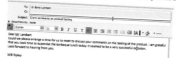

Photographs to show testing product:

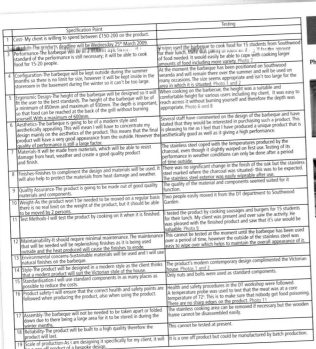

	Specification Point	Testing
1	Cost- My client is willing to spend between £150-200 on the product.	
2	Schedule-The product deadline will be Wednesday 25th March 2009.	
3	Performance-The barbeque will be of a modern style however a high standard of the performance is still necessary; it will be able to cook food for 15-20 people.	I mainly used the barbeque to cook food for 15 students from Southwood for their lunch, there was plenty of room and for the amount of food needed. It would easily be able to cope with cooking larger amounts of food including more variety. Photo 7
4	Configuration-The barbeque will be kept outside during the summer months so there is no limit for size, however it will be kept inside in the storeroom in the basement during the winter so it can't be too large.	At the moment the barbeque has been positioned on Southwood veranda and will remain there over the summer and will be used on many occasions. The size seems appropriate and isn't too large for the area in which it is situated. Photo 1 and 2
5	Ergonomic Design-The height of the barbeque will be designed so it will fit the user to the best standards. The height of the barbeque will be of a minimum of 850mm and maximum of 930mm. The depth is important, so that food can be reached at the back of the grill without burning yourself. With a maximum of 600mm.	When cooking on the barbecue, the height was a suitable and comfortable height for various users including my client. It was easy to reach across it without burning yourself and therefore the depth was appropriate. Photo 6 and 8
6	Aesthetics-The barbeque is going to be of a modern style and aesthetically appealing. This will mean I will have to concentrate my design mainly on the aesthetics of the product. This means that the final product will have a very good appearance from the outside. However the quality of performance is still a large factor.	Several staff have commented on the design of the barbecue and have stated that they would be interested in purchasing such a product. This is pleasing to me as I feel that I have produced a unique product that is aesthetically good as well as it giving a high performance.
7	Materials-It will be made from materials, which will be able to resist damage from heat, weather and create a good quality product and finish.	The stainless steel coped with the temperatures produced by the charcoal, even though it slightly warped on first use. Testing of its performance in weather conditions can only be done after a period of time outside.
8	Finishes-Finishes to compliment the design and materials will be used. It will also help to protect the materials from heat damage and weather.	There was no significant change in the finish of the oak but the stainless steel marked where the charcoal was situated- this was to be expected. The stainless steel exterior was easily wipeable after use.
9	Quality Assurance-The product is going to be made out of good quality materials and components.	The quality of the material and components seemed suited for it function.
10	Weight-As the product won't be needed to be moved on a regular basis there is no real limit on the weight of the product, but it should be able to be moved by 2 persons.	Two people easily moved it from the DT department to Southwood Garden.
11	Test Methods-I will test the product by cooking on it when it is finished.	I tested the product by cooking sausages and burgers for 15 students for their lunch. My client was present and saw the activity. He was pleased with the finished product and saw that it's use would be valuable. Photo 9
12	Maintainability-It should require minimal maintenance. The maintenance that will be needed will be replenishing finishes as it is being used outside and the heat produced will cause the finishes to erode.	This cannot be tested at the moment until the barbeque has been used over a period of time, however the outside of the stainless steel was easy to wipe over which helps to maintain the overall appearance of it.
13	Environmental concerns-Sustainable materials will be used and I will use natural finishes on the barbeque.	
14	Style-The product will be designed in a modern style as the client thinks that a modern product will suit the Victorian style of the house.	The product's modern contemporary design complimented the Victorian house. Photos 1 and 2
15	Standardisation-I will use standard components in as many places as possible to reduce the costs.	Only nuts and bolts were used as standard components.
16	Product safety-I will ensure that the correct health and safety points are followed when producing the product, also when using the product.	Health and safety procedures in the DT workshop were followed. A temperature probe was used to test that the meat was at a core temperature of 72°. This is to make sure that nobody got food poisoning. There are no sharp edges on the product. Photo 11
17	Assembly-The barbeque will not be needed to be taken apart or folded down due to there being a large area for it to be stored in during the winter months.	The stainless cooking area can be removed if necessary but the wooden frame cannot be disassembled easily.
18	Reliability-The product will be built to a high quality therefore the product will last.	This cannot be tested at present.
19	Scale of production-As I am designing it specifically for my client, it will be a one off product of a bespoke design.	It is a one off product but could be manufactured by batch production.

Moderator's comments

Some very good testing has been carried out by the student and most of the specification points have been tested or evaluated, although there is no reference here to sustainability. Some measurable tests were carried out, such as using a probe to measure the temperature of the food as it was being cooked to ensure that it was safe to eat. An interview was carried out with the person the barbeque was designed for and a small questionnaire was used to get the opinions of some of the users. Good photographic evidence is provided to show the product being tested and to show its fitness for purpose.

The work of the student here is judged as:	
Needs improvement	
Average	
Good	
Very good	✔

examzone

Welcome to ExamZone! Revising for your exams can be a daunting prospect. In this section of the book we'll take you through the best way of revising for your exams, step by step, to ensure you get the best results that you can achieve.

Zone In!

Have you ever become so absorbed in a task that it suddenly feels entirely natural? This is a feeling familiar to many athletes and performers: it's a feeling of being 'in the zone' that helps you focus and achieve your best. Here are our top tips for getting in the zone with your revision.

UNDERSTAND IT

Understand the exam process and what revision you need to do. This will give you confidence and help you to put things into proportion. These pages are a good place to find some starting pointers for performing well at exams.

BUILD CONFIDENCE

Use your revision time not just to revise the information you need to know, but also to practise the skills you need for the examination. Try answering questions in timed conditions so that you're more prepared for writing answers in the exam. The more prepared you are, the more confident you will feel on exam day.

DEAL WITH DISTRACTIONS

Think about the issues in your life that may interfere with revision. Write them all down. Think about how you can deal with each so they don't affect your revision. For example, revise in a room without a television, but plan breaks in your revision so that you can watch your favourite programmes. Be really honest with yourself about this – lots of students confuse time spent in their room with time revising. It's not at all the same thing if you've taken a look at Facebook every few minutes or taken mini-breaks to send that vital text message.

FRIENDS AND FAMILY

Make sure that they know when you want to revise, and even share your revision plan with them. Help them to understand that you must not get distracted. Set aside quality time with them, when you aren't revising or worrying about what you should be doing.

GET ORGANISED

If your notes, papers and books are in a mess you will find it difficult to start your revision. It is well worth spending a day organising your file notes with section dividers and ensuring that everything is in the right place. When you have a neat set of papers, turn your attention to organising your revision location. If this is your bedroom, make sure that you have a clean and organised area to revise in.

KEEP HEALTHY

During revision and exam time, make sure you eat well and exercise, and get enough sleep. If your body is not in the right state, your mind won't be either – and staying up late to cram the night before the exam is likely to leave you too tired to do your best.

Planning Zone

The key to success in exams and revision often lies in the right planning. Knowing what you need to do and when you need to do it is your best path to a stress-free experience. Here are some top tips in creating a great personal revision plan.

MY PLAN

1. Know when your exam is
Find out your exam dates. Go to www.edexcel.com/iwantto/Pages/dates.aspx to find all final exam dates, and check with your tutor. This will enable you to start planning your revision with the end date in mind.

2. Know your strengths and weaknesses
At the end of the chapter that you are studying, complete the 'You should know' checklist. Highlight the areas that you feel less confident on and allocate extra time to spend revising them.

3. Personalise your revision
This will help you to plan your personal revision effectively by putting a little more time into your weaker areas. Use your mock examination results and/or any further tests that are available to you as a check on your self-assessment.

4. Set your goals
Once you know your areas of strength and weakness you will be ready to set your daily and weekly goals.

5. Divide up your time and plan ahead
Draw up a calendar, or list all the dates, from when you can start your revision through to your exams.

6. Know what you're doing
Break your revision down into smaller sections. This will make it more manageable and less daunting. You might do this by referring to the Edexcel GCSE specification, to the chapter objectives, or to the headings within each chapter.

7. Link it together
Consider how topics interrelate. For example, when you are revising finishing techniques, it would be sensible to cross-reference this to other parts of your work such as the aesthetic properties of woods, metals and polymers.

EXAM DAY!

8. Break it up
Revise one small section at a time, but ensure you give more time to topics that you have identified weaknesses in.

9. Be realistic
Be realistic about how much time you can devote to your revision, but also make sure you put in enough time. Give yourself regular breaks or different activities to give your life some variety. Revision need not be a prison sentence!

10. Check your progress
Make sure you allow time for assessing progress against your initial self-assessment. Measuring progress will allow you to see and celebrate your improvement, and these little victories will build your confidence for the final exam

Finally – stick to your plan!

Know Zone

Remember that different people learn in different ways. Some remember visually and therefore might want to think about using diagrams and other drawings for their revision; others remember better through sound or through writing things out. Think about what works best for you by trying out some of the techniques below.

REVISION TECHNIQUES

Highlighting: work through your notes and highlight the important terms, ideas and explanations so that you start to filter out what you need to revise.

Key terms: look at the key terms highlighted in bold in each chapter. Try to write down a concise definition for this term. Now check your definition against the glossary definition on p173.

Summaries: writing a summary of the information in a chapter can be a useful way of making sure you've understood it. But don't just copy it all out. Try to reduce each paragraph to a couple of sentences. Then try to reduce the couple of sentences to a few words!

Concept maps: if you're a visual learner, you may find it easier to take in information by representing it visually. Draw concept maps or other diagrams. These are particularly good at showing links. For example, you could create a concept map which shows how to learn about sustainability.

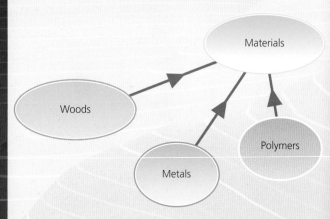

Mnemonics: this is when you take the first letter of a series of words you want to remember and then make a new word or sentence.

Index cards: Write important properties, definitions and processes on index cards and then test yourself.

Quizzes: Learning facts can be dull. Why not make a quiz out of it? Set a friend 20 questions to answer. Make up multiple-choice questions. You might even make up your own exam questions and see if your friend can answer them!

And then when you are ready:

Practice questions: go back through all the ResultsPlus features with questions to see if you can answer them (without cheating!). Try writing out some of your answers in timed conditions so that you're used to the amount of time you'll have to answer each type of question in the exam. Then, check the guidance for each one and try to mark your answer.

Use the list.

Chapter 1: p 12, 14, 20, 25 and 31

Chapter 2: p 37, 38 and 43

Chapter 3: p 48, 50, 52, 55, 58, 61, 65, and 72

Chapter 6: p 103, and 109

Chapter 7: p 113

Topic 1.1: Woods

RECALL: You should know which woods are hardwoods, softwoods and manufactured boards.

TRY IT: Working with a friend or partner, call out a specific wood type and say whether it is a hardwood, softwood or manufactured board. Take it in turns to do this. This is good revision for the multiple-choice questions.

APPLY: You need to be able to describe the aesthetic, functional and mechanical properties of woods when applied to an unfamiliar product or design question.

Examiner's tip

Take two sets of cards. Write the names of different woods on the first set, and write hardwood, softwood or manufactured board on the second set. Put the two sets of cards in separate piles and match them up.

Topic 1.2: Metals

RECALL: You should know which metals are ferrous and which are non-ferrous. You should also know the definitions of ductility, malleability, hardness, toughness, elasticity and strength in tension, compression and shear. You should know the composition of mild steel and brass.

TRY IT: List each material and add some notes about their advantages, disadvantages and properties.

APPLY: You need to be able to describe the aesthetic, functional and mechanical properties of metals when applied to an unfamiliar product or design question.

Examiner's tip

Take two sets of cards. Write the definitions of ductility, malleability, hardness, toughness, elasticity, strength (tension, compression and shear) on the first set, and the properties on the second set. Put the two sets of cards in separate piles and try to match the correct property to each definition.

Topic 1.3: Polymers

RECALL: You should know about thermoplastics and thermosetting plastics. You should also know the definitions of plasticity and durability.

TRY IT: List each material and add some notes about their advantages, disadvantages and properties.

APPLY: Be able to describe the aesthetic, functional and mechanical properties of polymers when applied to an unfamiliar product or design question.

Examiner's tip

Take two sets of cards. Write the definitions of plasticity and durability on the first set, and the properties on the second set. Put the two sets of cards in separate piles and try to match the correct property to each definition.

Take another two sets of cards and repeat the exercise, this time writing the names of specific polymers on the first set and either thermoplastic or thermosetting plastic on the second set.

Topic 1.4: Composites

RECALL: You should know about the uses of composites when manufacturing products.

TRY IT: Working with a friend or partner, take it in turns to state an advantage or disadvantage of a composite.

APPLY: Be able to describe the structural, functional and mechanical properties of composites when applied to an unfamiliar product.

Topic 1.5: Modern and smart materials

RECALL: You should know about the uses of modern and smart materials in manufacturing products.

TRY IT: Working with a friend or partner, take it in turns to state an advantage or disadvantage of a modern or smart material.

APPLY: Be able to describe the uses, advantages and disadvantages of modern and smart materials when applied to an unfamiliar product or design question.

Examiner's tip

Make sure you know what a smart material is. All of the following are smart materials: shape memory alloys (SMA), photochromic paint, reactive glass and carbon nanotubes.

Topic 2.1: Marking out and measuring

RECALL: You should know about the following tools: rules, squares, gauges, scribers, punches, templates and micrometers.

TRY IT: Create some picture cards showing the tools and take it in turns with a partner to name each tool and describe its use.

APPLY: Be able to recognise each of the listed tools and give the advantages and disadvantages of its use when marking out and measuring during the manufacture of an unfamiliar product.

Examiner's tip

Make sure that you can name all the tools listed and describe their uses. This will be very useful when it comes to the section on your exam paper where you are asked to name and describe the uses of a particular tool.

Topic 2.2: Wasting

RECALL: You should know about the uses of the following tools when removing material during the manufacture of products: saws, planes, chisels, files, drills, abrading tools.

TRY IT: Create some picture cards showing the tools and take it in turns with a partner to name each tool and describe its use.

APPLY: Be able to describe the uses, advantages and disadvantages of the tools listed above.

Examiner's tip

Make sure that you can name all the tools listed and describe their use. This will be very useful when it comes to the section on your exam paper where you are asked to name and describe the uses of a particular tool.

Topic 3.1: Scale of production

RECALL: You should know about one-off, batch and mass production.

TRY IT: Explain the advantages and disadvantages of each type of production to a friend.

APPLY: Be able to describe and explain why certain products are made as one-offs, in batches or as mass-produced items.

Examiner's tip
Make sure that you can give some examples of products that fall into each of the three levels of production.

Topic 3.2: Materials processing and forming

RECALL: You should know about the following methods of production: sand casting, drilling, turning (wood and metal), blow moulding, injection moulding, vacuum forming, extrusion (plastic and metal) and wood laminating.

TRY IT: Working with a friend or partner, take it in turns to describe one of the processes listed above.

APPLY: Be able to describe the uses, advantages and disadvantages of the processes listed above when applied to an unfamiliar product or design question.

Examiner's tip
Make sure that you can give some examples of products that are made by each of the listed processes.

Topic 3.3: Joining methods

RECALL: You should know about the following methods of joining: tapping and threading, nuts bolts and washers, screws, knock-down fittings, nails, halving joints, butt joints, rebate joints, housing joints, mortise and tenon joints, dowel joints, soft soldering, brazing, welding and rivets.

TRY IT: Working with a friend or partner, take it in turns to describe one of the joining methods listed above.

APPLY: Be able to describe the uses, advantages and disadvantages of joining methods when applied to an unfamiliar product.

Examiner's tip
Make sure that you can name all the joints listed and describe their use.

Topic 3.4: Adhesives

RECALL: You should know about the following adhesives: polyvinyl acetate (PVA), contact adhesive, epoxy resin and Tensol® cement.

TRY IT: Explain the advantages and disadvantages of each of the adhesives to a friend.

APPLY: Be able to select an adhesive for an unfamiliar situation.

Examiner's tip

Make sure you know what materials can be joined with each adhesive.

Topic 3.5: Heat treatment

RECALL: You should know about the following methods of heat treatment: hardening and tempering, annealing and case hardening.

TRY IT: Working with a friend or partner, take it in turns to describe one of the processes listed above.

APPLY: Be able to describe the uses, advantages and disadvantages of the processes listed above.

Examiner's tip

Make sure that you can give some examples of products that have been heat-treated.

Topic 3.6: Finishing techniques

RECALL: You should know about the following finishing techniques: varnish, wax polish, stain, paint, plastic dip-coating and electroplating.

TRY IT: Explain the advantages and disadvantages of each of the finishing techniques listed above to a friend.

APPLY: Be able to describe and explain why certain products are finished using a particular method.

Examiner's tip

Make sure that you can give some examples of products that are finished with each of the listed finishes.

Topic 3.7: Manufacturing processes for batch production

RECALL: You should know about the following methods of production when producing products or components in batches: jigs and patterns.

TRY IT: Working with a friend or partner, take it in turns to describe how jigs and patterns are used in the manufacture of batch produced products or components.

APPLY: Be able to describe the advantages and disadvantages of jigs and patterns.

Examiner's tip
Make sure you know the difference between a jig and a pattern.

Topic 3.8: Health and safety

RECALL: Be able to identify workshop hazards and precautions.

TRY IT: Working with a friend or partner, take it in turns to describe safe working practices.

APPLY: You should know how to describe safe working practices.

Examiner's tip
Make sure you know what precautions to take when using machinery.

Topic 4.1: Specification criteria

RECALL: You should be able to use the following criteria when analysing products: form, function, user requirements, performance requirements, material and component requirements, scale of production and cost, sustainability.

TRY IT: Working with a friend or partner, take it in turns to analyse a product using the criteria above.

APPLY: Be able to take into account the criteria listed when analysing unfamiliar products in an exam.

Examiner's tip

Make sure you know what is meant by each of the above criteria.

Topic 4.2: Materials and components

RECALL: You should know how to identify materials and/or components used in the manufacture of an unfamiliar product

TRY IT: Working with a friend, discuss the possible materials and/or components used in the manufacture of an unfamiliar product.

APPLY: Be able to describe the properties and qualities of materials and components, including their advantages and disadvantages.

Examiner's tip

You only need to know about the materials and components listed in Topic 1.

Topic 4.3: Manufacturing processes

RECALL: You should be able to identify the processes involved in the manufacture of unfamiliar products.

TRY IT: With a friend identify the stages of the manufacturing processes of some household items.

APPLY: Be able to describe the advantages and disadvantages of the manufacturing processes listed in topic 3.

Examiner's tip

You only need to know about the manufacturing processes listed in Topic 3.

Topic 5.1: Specification criteria

RECALL: You should be able to use the following criteria when designing products: form, function, user requirements, performance requirements, material and component requirements, scale of production and cost, sustainability.

TRY IT: Try designing a new product using the criteria above.

APPLY: Be able to take into account the above criteria when designing unfamiliar products in an exam.

Examiner's tip
Make sure you know what is meant by each of the criteria.

Topic 5.2: Designing skills

RECALL: You should be able to use your design skills in order to produce solutions to unfamiliar design briefs and list of specification criteria.

TRY IT: Practise your drawing skills to present solutions in about eight minutes.

APPLY: Be able to use clear communication to show design intentions using notes and/or sketches.

Examiner's tip
Clearly present your drawings and annotation in the examination.

Topic 5.3: Application of knowledge and understanding

RECALL: You should be able to apply your knowledge and understanding of a wide range of materials and/or components and manufacturing processes when designing unfamiliar products.

TRY IT: Be able to identify the properties of materials and/or components.

APPLY: Be able to describe the advantages and disadvantages and justify the choice of materials/components and manufacturing processes.

Examiner's tip
Make sure you fully annotate your design work in the examination.

Topic 6.1: Information and communication technology (ICT)

RECALL: You should know about electronic communications using email, electronic point of sale (EPOS) in product retail and manufacture, and internet marketing and sales.

TRY IT: Explain to a friend how designers, manufacturers, retailers and consumers can communicate electronically using email.

APPLY: Be able to describe the uses, advantages and disadvantages of ICT in the design, development and marketing of products.

Examiner's tip
Make sure you know what EPOS stands for.

.

Topic 6.2: Digital media and new technology

RECALL: You should know about the transfer of data using Bluetooth® wireless personal area networks and videoconferencing.

TRY IT: Working with a friend or partner, take it in turns to describe one of the communication methods listed above.

APPLY: Be able to describe the uses, advantages and disadvantages of transferring data using Bluetooth® and videoconferencing.

Topic 6.3: Computer-aided design/computer-aided manufacturing (CAD/CAM)

RECALL: You should know about the following: virtual modelling and testing, laser cutting, computer numerically controlled (CNC) milling and turning, rapid prototyping.

TRY IT: Be able to explain the advantages and disadvantages of each of the CAD/CAM processes to a friend.

APPLY: Be able to describe and explain why certain processes are used.

Examiner's tip
Make sure that you can give some examples of products manufactured using the processes listed.

Topic 7.1: Minimising waste production

RECALL: You should know about the 4 Rs: reducing materials and energy, reusing materials and products, recovering energy from waste and recycling.

TRY IT: Explain how waste can be minimised using the 4Rs to a friend.

APPLY: Be able to describe the uses, advantages and disadvantages of the 4Rs.

Examiner's tip
Make sure that you can give some examples of reused, recovered and recycled products.

Topic 7.2: Renewable sources of energy

RECALL: You should know about wind energy, solar energy and how biomass is converted into biofuels for transportation.

TRY IT: Working with a friend or partner, take it in turns to describe one of the renewable sources of energy listed above.

APPLY: Be able to describe the uses, advantages and disadvantages of these renewable energy sources.

Examiner's tip
Make sure you understand what is meant by the term 'renewable resources'.

Topic 7.3: Climate change

RECALL: You should be aware of the responsibilities of developed countries in minimising global warming, including how the Kyoto Protocol seeks to reduce greenhouse gas emissions.

TRY IT: Working with a friend or partner, discuss the responsibilities of developed countries in minimising the impact of industrialisation on global warming and climate change.

APPLY: Be able to describe and explain why such changes are needed.

Examiner's tip
Make sure you learn about the responsibilities of developed countries in relation to climate change.

Topic 8.1: Moral, social and cultural issues

RECALL: You should know about built-in obsolescence, offshore manufacturing of mass-produced products in developing countries, and the importance of respecting different cultures when designing and manufacturing products..

TRY IT: Working with a friend or partner, discuss value issues when designing and manufacturing products.

APPLY: Be able to describe the uses, advantages and disadvantages of built-in obsolescence and offshore manufacturing.

Examiner's tip
Make sure you understand the term 'built-in obsolescence'.

Don't Panic Zone

Once you have completed your revision in your plan, you'll be coming closer and closer to the big day. Many learners find this the most stressful time and tend to go into panic-mode, either working long hours without really giving their brain a chance to absorb information, or giving up and staring blankly at the wall. Follow these tips to ensure that you don't panic at the last minute.

TOP TIPS

1. Test yourself by analysing the materials and manufacturing processes used in the manufacture of new products.

2. Look over past exam papers and their mark schemes. Look carefully at what the mark schemes are expecting of candidates in relation to the question.

3. Do as many practice questions as you can to improve your technique, help manage your time and build confidence in dealing with different questions.

4. Write down a handful of the most difficult bits of information for each chapter that you have studied. At the last minute, focus on learning these.

5. Relax the night before your exam – last-minute revision for several hours rarely has much additional benefit. Your brain needs to be rested and relaxed to perform at its best.

6. Remember the purpose of the exam – it's for you to show the examiner what you have learnt.

LAST MINUTE LEARNING TIPS FOR DESIGN TECHNOLOGY

● Remember that an intelligent guess is better than nothing. If you can't think of a property then take a guess – you cannot lose marks.

● Know your materials and properties – don't go into the exam unclear about basic definitions: ductility, malleability, hardness, toughness and elasticity. Check out the glossary.

● Many exam questions ask you to apply your knowledge and understanding to unfamiliar products. Make sure that you revise manufacturing processes and properties that apply to all materials.

Exam Zone

Here is some guidance on what to expect in the exam itself: what the questions will be like and what the paper will look like.

UNIT	% OF OVERALL GCSE	MARKS	DESCRIPTION	KNOWLEDGE AND SKILLS
Unit 1: Creative Design and Make Activities	60	100 Design: 50 Make: 50	• You can choose one design activity and a separate make activity which are not related, or you can choose one combined task. • It is recommended that you take no longer than 40 hours to complete the whole task. • The task is assessed internally and externally moderated. • There is an opportunity to re-take this unit once. • The project can be submitted at the end of Year 10.	The following skills will be tested: • design creatively • make products • apply systems and control, CAD/CAM. digital media and new technologies where appropriate • analyse and evaluate processes and products.
Unit 2: Knowledge and Understanding of RMT	40	80	• There is a single tier paper. • The exam is 90 minutes. • You must answer all the questions. • There are 10 multiple choice questions at the start of the paper. • There will be a design question and a product analysis question. • The marks for each question or part question are shown in brackets at the end of each question or part question. • Extended writing questions examining the quality of your written communication are marked with an *. • The paper will be externally assessed. • There is an opportunity to retake the exam once. • The exam can be taken at the end of Year 10.	The following are tested: • materials and components • tools and equipmet • industrial and commercial processes • analysing products • designing products • technology • sustainability • ethical design and manufacture.

ASSESSMENT OBJECTIVES

The questions you will be asked are designed to examine the following aspects of deign technology. These are known as assessment objectives (AO). There are three AOs.

AO1 Recall, select and communicate knowledge and understanding in design and technology, including its wider effects.

AO2 Apply new knowledge, understanding and skills in a variety of contexts and in designing and making products.

AO3 Analyse and evaluate products, including their design and production.

THE TYPES OF QUESTION THAT YOU CAN EXPECT IN YOUR EXAM

The exam papers are designed so that the opening part of each question is the easiest part. They become more difficult as you move through the question. The level of difficulty is controlled by the command word and content required in your answers.

There are four different types of question:

MCQ Multiple-choice question.

Short Single-word answers or responses involving a simple phrase or statement.

Open Free response questions that involve a limited amount of continuous prose.

Long Free response questions where you have the opportunity for extended writing. These allow the quality of your written communication to be assessed.

UNDERSTANDING THE LANGUAGE OF THE EXAM PAPER

It is vital that you know what 'command' words ask you to do. Common errors are:

1. not fully *describing* or *explaining* an answer
2. adding a *description* or *explanation* when you do not have to.

Command word	Marks awarded	Description
Give/state/name	(1 mark)	This type of question will usually appear at the beginning of the paper or question part and is designed to ease you into the question with a simple statement or short phrase.
Describe/outline	(2+ marks)	This type of question is quite straightforward. It asks you simply to describe something in detail. Some questions may also ask you to use notes and sketches, so you can gain marks with the use of a clearly labelled sketch.
Explain/justify	(2+ marks)	This type of question asks you to respond in detail to the question – no short phrases will be acceptable here. Instead, you will have to make a valid point and develop/justify it to get full marks.
Evaluate/discuss/ compare	(4+ marks)	This type of question is designed to stretch and challenge you. It will always carry the most marks because you are required to make a well-balanced argument, usually involving both advantages and disadvantages.

Meet the exam paper

This section shows you what the exam paper looks like. Check that you understand each part. Now is a good opportunity to ask your tutor about anything that you are not sure of here.

Print your surname here, and your other names in the next box. This is an additional safeguard to ensure that the exam board awards the marks to the right candidate.

Here you fill in your personal exam number. Take care when writing it down because the number is important to the exam board when writing your score.

Ensure that you understand exactly how long the examination will last, and plan your time accordingly.

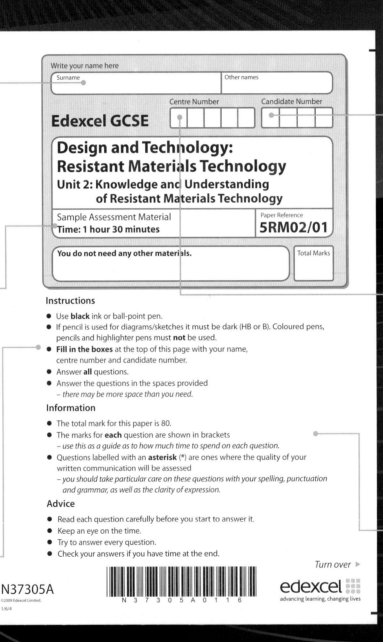

Write your name here

Surname

Other names

Centre Number   Candidate Number

Edexcel GCSE

Design and Technology: Resistant Materials Technology
Unit 2: Knowledge and Understanding of Resistant Materials Technology

Sample Assessment Material
Time: 1 hour 30 minutes

Paper Reference
5RM02/01

You do not need any other materials.

Total Marks

Instructions

● Use **black** ink or ball-point pen.
● If pencil is used for diagrams/sketches it must be dark (HB or B). Coloured pens, pencils and highlighter pens must **not** be used.
● **Fill in the boxes** at the top of this page with your name, centre number and candidate number.
● Answer **all** questions.
● Answer the questions in the spaces provided
  – there may be more space than you need.

Information

● The total mark for this paper is 80.
● The marks for **each** question are shown in brackets
  – use this as a guide as to how much time to spend on each question.
● Questions labelled with an **asterisk** (*) are ones where the quality of your written communication will be assessed
  – you should take particular care on these questions with your spelling, punctuation and grammar, as well as the clarity of expression.

Advice

● Read each question carefully before you start to answer it.
● Keep an eye on the time.
● Try to answer every question.
● Check your answers if you have time at the end.

Turn over ▶

N37305A
©2009 Edexcel Limited.
1/6/4

N 3 7 3 0 5 A 0 1 1 6

edexcel
advancing learning, changing lives

Here you fill in your school's centre number. You will be given this by your teacher on the day of your exam.

Note that the quality of your written communication will also be marked. Take particular care to present your thoughts and work at the highest standard you can for maximum marks.

Ensure that you read the instructions carefully and that you understand exactly which questions from which sections you should attempt.

Answer ALL the questions.

For each question 1 to 10, choose an answer A, B, C or D. Put a cross in the box indicating the answer you have chosen ☒. If you change your mind about an answer, put a line through the box ☒ and then mark your new answer with a cross ☒.

1   Which **one** of the following is used as a measuring tool?

   ☐ **A**   Try square

   ☐ **B**   Centre punch

   ☐ **C**   Micrometer

   ☐ **D**   Scriber

   (Total for Question 1 = 1 mark)

2   The frame below is joined at the corner with what type of joint?

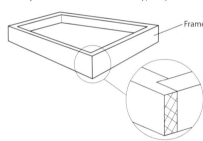

Frame

   ☐ **A**   Knock down fitting

   ☐ **B**   Mortise and tenon

   ☐ **C**   Dovetail

   ☐ **D**   Rebate

   (Total for Question 2 = 1 mark)

3   Which type of saw is best suited to cutting a curve in a piece of plywood?

   ☐ **A**   Hacksaw

   ☐ **B**   Coping saw

   ☐ **C**   Tenon saw

   ☐ **D**   Back saw

   (Total for Question 3 = 1 mark)

2

N 3 7 3 0 5 A 0 2 1 6

Read the instructions each time – they are there to provide guidance.

Pay attention to any text highlighted in bold. It is highlighted to alert you to important information, so be sure to read it and take note!

The marks for each question are shown on the right-hand side of the page. Make sure that you note how many marks a question is worth as this will give you an idea of how long to spend on that question.

Zone Out

Well done, you have finished your exam. So, what now? This section provides answers to the most common questions students have about what happens after they complete their exams.

About your grades

Whether you've done better than, worse than or just as you expected, your grades are the final measure of your performance on your course and in the exams.

When will my results be published?

Results for summer examinations are issued in August. January exam results are issued in March.

Can I get my results online?

Visit www.resultsplusdirect.co.uk, where you will find detailed student results information including the 'Edexcel Gradeometer', which demonstrates how close you were to the nearest grade boundary. Students can only gain their results online if their centre gives them permission to do so.

I haven't done as well as I expected. What can I do now?

First of all, talk to your teacher. After all the teaching that you have had, tests and internal examinations, he/she is the person who best knows what grade you are capable of achieving. Take your results slip to your teacher, and go through the information on it in detail. If you both think that there is something wrong with the result, the school or college can apply to see your completed examination paper and then, if necessary, ask for a re-mark immediately. The original mark can be confirmed or lowered, as well as raised, as a result of a re-mark.

If I am not happy with my grade, can I resit a unit?

Yes, you are able to resit each unit once before claiming certification for the qualification. The best available result for each contributing unit will count towards your final grade.

What can I do with a GCSE in Design Technology?

Design Technology is well-known as a subject that links to all other subjects of the curriculum, so a GCSE in Design Technology is a stepping stone to a whole range of opportunities. A good grade will help you to move on to AS, to A Level or to a BTEC course. If you have enjoyed the subject you may want to go on to study A level Product Design: RMT for a further two years, along with other A level subjects such as Maths and the Sciences. Creative students usually study one or more of the creative subjects including A level Art and Design, Media and/or Film, BTEC National Diplomas in Art and Design or Media and the 14–19 Diploma in Creative and Media. Of course, if post-16 study is not for you, employers value this GCSE RMT qualification as it develops creative, technical and transferable skills. The skills that you develop can lead you to employment opportunities in the creative industries, construction, engineering and ICT.

Glossary

Aesthetics: How we respond to visual appearance

Alloys: A group of metals made by combining two or more metallic elements

Annealing: The process of heating metals and allowing them to cool slowly in order to remove internal stresses

Annual ring: Growth ring produced each year as a tree grows

Batch production: A method of production where a number of components are made

Bespoke: A product that has been made to order

Bevel: A sloping surface or edge

Billet: Small bar of metal that needs further processing

Biofuel: A fuel derived from living organic mater

Biomass: Organic matter that can be used for fuel

Bluetooth®: Concerning the use of short-range wireless communications for mobile phones, computers, etc.

Bowing: Becoming bent along the length of the piece of wood

Built-in obsolescence: Where products are designed to stop working or become redundant after a set period

Cascamite: A type of adhesive supplied as a powder

Case-hardening: To create a hard surface on the outside of a metal surface by carburising

CNC: Computer numerically controlled; used of a range of machinery controlled by computers

Composites: A group of materials made from a combination of two or more different materials, in layers or as a mixture

Compressive strength: The ability of a material to withstand being squashed

Coniferous: Associated with a cone-bearing (softwood) tree

Cross-links: Chains of molecules that link together with bonds to form a rigid structure

Cupping: Becoming rolled across the width of the plank

Datum: A flat face or straight edge from which all measurements are taken

Deciduous: Losing leaves in winter (with a few exceptions)

Deforming: A process that allows materials to change shape without changing state, e.g. vacuum forming

Die stock: A tool used to hold a split die

Dimensional stability: Having dimensions that will not change when subjected to extreme environmental conditions

Dovetail: A type of woodworking joint

Draft angle: The tapered side of a mould

Ductile: Of a material, capable of being drawn or stretched into thinner, smaller sections

Durability: The ability of a material to withstand wear, pressure or damage

Elasticity: The ability of a material to return to its original shape once a deforming force has been removed

Electrical insulators: A group of materials that will not allow a current to pass through them

Electrolysis: Chemical decomposition produced by passing an electric current thorough a conducting solution

Engineer's blue: A liquid to coat metals before marking out

EPOS: Electronic point of sale

Extruded: Squeezed through; the process is used extensively for plastics and metals as it forms uniform cross-sections

Face edge: The surface at right angles to the face side

Face side: The side chosen to take measurements from

Faceplate: An attachment that fixes onto the outside of a lathe so that products such as fruit bowls can be turned

Felled: Of a tree, cut down in order to produce timber for use

Ferrous: Containing iron

Fibrous: Made up of tiny, thread-like fibres

Flux: A material added around a joint when brazing to help solder flow and prevent build up of excess surface oxides

Form: Why a product is shaped or styled as it is

Former: A base on which to build up thin layers of a material

Function: What the purpose of the product is

Galvanising: Applying a protective layer of zinc to an iron or steel surface

Hardwood: Wood from broadleaved trees

Hardening: The heat-treatment process of making a metal harder by applying heat and cooling rapidly

Hardness: The ability to withstand abrasive wear and indentation

High tensile strength: The ability of a material to withstand being pulled apart or stretched

Knots: Hard patches formed where a new branch grows from the main trunk of a tree or from another branch

Kyoto Protocol: An international agreement that set out to reduce greenhouse gas emissions globally

Laser cutting: Using high-powered lasers to cut materials

Malleable: Of a material, capable of being deformed by compression without tearing or cracking

Mass production: The production of a component or product in large numbers

Molecular structure: The way molecules are arranged within a material or element

Non-ferrous: Containing no iron

Offshore manufacturing: The practice of large manufacturing companies and industries relocating their businesses to take advantage of lower costs

One-off production: Making a product as a single item, such as a bridge or a football stadium

Opaque: Not able to be seen through

Organic material: Derived from living matter

Parison: A tube-like shape made of plastic material prepared for use in blow moulding

Performance requirements: The technical considerations that must be achieved within the product

Photovoltaic cell: An electronic device that can be used to generate electricity from the sun's energy

Plastic memory: The ability of a plastic to return to its original shape and form when softened; this only applies to thermoplastics if their molecular structure has not been damaged

Plasticisers: Additives incorporated into polymers to change their properties

Plasticity: The ability of a material to be changed permanently without cracking or breaking

Polymer: A plastic or synthetic resin made up from molecules bonded together

Rapid prototyping: A process that automatically creates physical objects, by building up thin layers or using processes such as stereolithography

Recover: Get energy from waste materials

Recycle: Process used materials and products into new materials or products

Reduce: Lower the amount of energy or material used in the manufacture of products

Reuse: Use a product or material more than once

Reforming: A process that involves a change in state of the material being processed, e.g. casting

Router: A hand-held tool capable of holding different sized and shaped bits to cut different slots or profiles

Seasoning: The process of drying to reduce the moisture content of newly cut-down timber

Shear strength: The ability of a material or joint to withstand being slid apart

Softwood: Wood from a cone-bearing tree (conifer)

Split die: A tool used to cut an external screw thread

Splitting: When the end of a wooden plank splits as it dries out

Sprue: A channel through which metal or plastic can be poured or injected

Sustainable: Capable of being maintained at a certain level

Swarf: Small bits of waste material produced while cutting screw threads or when cutting on a centre lathe

Sweated: When two pieces of tinned metal are joined together

Tap: A tool used for cutting an internal screw thread

Tap wrench: A tool used to hold a tap

Tarnish: A film or stain that forms on an exposed surface, often leading to a change in colour or loss of lustre

Tempering: The heat-treatment process of removing excess brittleness once a component has been hardened

Tensile strength: The ability of a material to withstand being pulled apart

Thermoplastics: A group of plastics that repeatedly become soft on heating and harden on cooling

Thermosetting plastics: A group of plastics that set permanently when heated

Timber: Wood prepared for use in construction

Tinning: Applying a thin layer of solder to two surfaces to be joined by soldering

Toughness: The ability to withstand sudden and shock loading without fracture

User requirements: The qualities potential users want

UV: Ultraviolet, having a wavelength shorter than that of the violet end of the spectrum

Veneers: Thin layers of wood that are stuck to cheaper woods

Versatile: Capable of being used for or adapted to many different applications

Videoconference: A virtual meeting between people in different locations using television, video or computer

Virtual modelling: Creating models on a computer

Warping: Becoming bent or twisted out of shape

Index